FUNGI

Around Some Aquatic Bodies In Andhra Pradesh, India

***Vasant Rao* *,**
M.Sc., Ph.D
Professor of Botany (Retired)
Presently, Principal Investigator UGC, Major Research Project

C. Manoharachary,
M.Sc., Ph.D., D.Sc., FPSI., FBS., FAPAS
Professor of Botany
Presently, Principal Investigator UGC, DBT and MoENF, Major Research Projects

G. Suresh Kumar,
M.Sc., Ph.D

K. Subodh,
M.Sc., M.Tech., Ph.D

Department of Botany, Osmania University,
Hyderabad - 500 007. A.P.
INDIA

* *Supported by*

Department of Science & Technology (DST under USER'S)
Technology Bhavan, Meharauli Road
New Delhi, INDIA.

BSP BS Publications
A unit of **BSP Books Pvt., Ltd.**

4-4-309/316, Giriraj Lane, Sultan Bazar,
Hyderabad - 500 095
Phone : 040 - 23445605, 23445688

Funds channeled through
Society for Conservation and Management of Natural Resources
Aparajeeta Apartments, Ameerpet, Hyderabad - 500 029.

Published by

BSP BS Publications
A unit of **BSP Books Pvt., Ltd.**

4-4-309/316, Giriraj Lane, Sultan Bazar,
Hyderabad - 500 095
Phone : 040 - 23445605, 23445688
e-mail : info@bspbooks.net

ISBN: 978-93-52300-14-3 (HB)

Dedicated

to

Late Sri L. B. Deshpande

Laxman Rao Bhim Rao Deshpande, popularly known as LB and Lala at home, joined Osmania University where, by his devotion, dedication and human touch became the Controller of Examinations which itself reflects the confidence reposed in him by the university. Later, he became Registar of the University. As Registar, he introduced several reforms which got etched over the profile of this great University. He was a friend, philosopher and guide to many institutions of excellence. His life reflected simplicity, honesty and cheerful spirit to lead a team. He was great in action and thought. It is a pleasure to dedicate this volume viz., *"Fungi Around Some Aquatic Bodies in Andhra Pradesh, India"* to his memory as a testimony of profound respect, for he synthesised the administrative and academic spheres of the university.

Prof. Vasant Rao

Prof. C. Manoharachary

Dr. G. Suresh Kumar

Dr. K. Subodh

ASIAN AGRI-HISTORY FOUNDATION
(AAHF)

47 ICRISAT Colony-1
Brig. Sayeed Road
Secunderabad 500 009
India

Tel: +91 (40) 27755774
Fax: +91 (40) 27750630
E-mail: ylnahf@hd1.vsnl.net.in

Foreword

I am delighted to see the publication, Fungi Around Some Aquatic Bodies in Andhra Pradesh, India, authored by Drs. Vasant Rao, Manoharachary, Suresh Kumar, and Subodh, and which is based on their researches.

The Department of Botany of the Osmania University, Hyderabad has an excellent track record of research contributions over several decades. The present publication is yet another timely and potentially useful contribution.

Though the first half of the twentieth century produced many great mycologists in India, the second half saw their numbers dwindling. This was very unfortunate because the crop agriculture needs full support of mycologists, and there are currently too few mycologists to meet the needs. Mycology needs continued support.

The information presented in the book will be of great interest to those mycologists, plant pathologists, and others, who are working in the field of agriculture and biodiversity. Crop scientists researching irrigated agriculture will specially find the information useful.

I congratulate the authors for this valuable contribution to mycology and related disciplines.

Y L Nene
Chairman
Asian Agri-History Foundation, and
formerly Deputy Director General of
International Crops Research Institute
for the Semi-Arid Tropics (ICRISAT)
Hyderabad

Preface

Inspite of having a good network of R & D establishments, National and International Policy frame work, adequate natural and human resource wealth, advancements in Biotechnology and Information Technology, India's achievements in biodiversity and related areas of research did not receive much attention. Lack of co-ordination and integrated efforts, reluctance to accept and support Basic Science disciplines like Systematics, lack of collective vision and common agenda for tackling complex research and development, are seen as some of the inherent weaknesses of our R&D systems particularly with reference to Systematic Biology research. It is therefore, necessary to examine the problems of Systematic Biology research and suggest to evolve a national mechanism to coordinate all on-going and future research in Systematics and Taxonomy and to integrate it to meet the national goals of inventory, conservation, prospecting and sustainable use of biodiversity resources of our country.

With such background, Society for Conservation and Management of Natural Resources (Regd) was started in 1992 at Hyderabad by a group of scientists and enthusiasts in scientific studies, along with the excellent co-operation and dedication of Late Dr. Kaza Rama Rao of Zoological Survey of India (Founder Secretary), Late Prof. V. Vidyanath of Geography Department, Osmania University (Founder Vice President) and Prof. R. Seshagiri Rao of Botanical Survey of India and Andhra University (Founder Vice President) and others. Natural Resources of Andhra Pradesh and of India with special reference to plants including lower groups like Fungi etc., animals including reptile group, fisheries, molluscan group, meiofauna of sandy beaches, geographical data, District Floras of Andhra Pradesh including medicinal plants and mangrove areas, have been studied. Books for some of the subjects have been brought out by the Society.

With financial assistance of Dept. Science & Technology (DST), Botanical Survey of India (BSI), Council of Scientific and Industrial Research (CSIR) and a few other research organisations, the projects noted above have been worked out by various retired scientists. Recently during the year 2000, financial assistance from DST under "Retired Scientists USERS" programme has been very encouraging, leading to preparation of useful books and organising useful seminars on these subjects. This book presenting illustrative account of "Fungi around some aquatic bodies in Andhra Pradesh", along with other three books - "Common Reptiles of India - a Source Book", "Molluscan Fisheries of India" and "Meiofauna of Marine Beach Sand" are published under users programme of DST during the last 2 years (2000-2002) by our Society.

Andhra Pradesh is one of the largest States in India. It has several aquatic bodies in each district, covering various Mandals with numerous villages. The data presented in this volume is based on the study of 25-30 years by various workers including the authors of this book. Such presentation provides useful data to encourage subsequent workers in A.P.State and other parts of India to bring out more exhaustive details on different groups of fungi.

Fungi are beautiful to look at, but one must have scientific eye to appreciate their beauty as creation of the Nature. Today fungi are being used, or for that matter, as the tool in agriculture, pharmaceuticals, food and biotechnological industries and in several other scientific activities connected with human welfare. Fungi and other groups of plants and animals are intimately associated and form an integral part of human life. Fungi around water bodies play an important role in recycling of nutrients and they form the very basis of food chain and sustain aquatic and terrestrial life.

The present work is intended to create interest not only among the students of Mycology, Microbiology, Agriculture and other allied subjects but also among Environmentalists and Naturalists. It is also necessary to improve the standards of maintenance of suitable Herbaria, Cultures, and other methods of preservation to maintain their Genome which is under heavy pressure of disturbance by floods, fire and other aspects of human intervention along with their natural abodes.

Acknowledgment

Sincere thanks are due to Department of Science & Technology (DST), Govt. of India, New Delhi for financial support and giving chance to serve the scientific cause under USERS scheme (VR). Thanks are due to the President and Secretary of Society for Conservation and Management of Natural Resources (SCMNR), Dr. E.V. Muley, Department of Environment and Wild life, Dr. A.D. Diwan, IARI, New Delhi, Dr. M. S. Kodarkar, Prof. Bir Bahadur, Dr. Dev Rao, Dr. S.S.Kulkarni and Dr. A.Satyanarayna for their encouragement and valuable help.

Thanks are due to Prof. D.J. Bhat, University of Goa, Goa, and Prof. Alka Pande, ARI, Pune for their timely help.

Our thanks are due to Prof. G.S.de Hoog, Prof. Walter Gams, Prof. R.A. Samson, Prof. Van der Aa and Prof. Stalpers of CBS, Utrecht, Netherlands for their support and encouragement. We also thank Prof. S.J. Hughes, Dr. K.A. Seifert, Prof. W.B. Kendrick, Prof. T.R. Nagraj of Canada and Prof. B.C. Sutton, Dr. P.M. Kirk of IMI, U.K. Prof. Mohamad Bhai, Prof. A. Peerally, Mrs. Sunita, Madan Aba, Devanand Aba and Dev Rakhah of Mauritius for their best wishes and encouragement. Dr. A.K. Sarbhoy, Dr. P.N. Choudhary and Dr. D.K. Agarwal IARI, New Delhi are thanked for their help.

Thanks are due to Vice Chancellor, and Head Department of Botany, Osmania University authorities for providing facilities and Dept. of Forests, Govt. of AP for help in collection trips and Mr. Pranay Waghrey, Nalla Malai Foundation for their kind help.

Thanks are due to Mrs. Neha and Mr. Sunil V. K. for all their help in completing the book and to Mr. Nikhil Shah and Anil Shah of BS Publications, Hyderabad, for all co-operation and early publication.

Contents

I Introduction

I Introduction

Fungi are achlorophyllous, micro-macroscopic, holocarpic, eucarpic and eukaryotic living organisms leading saprophytic, biotrophic and symbiotic life. Some of the fungi live on living hosts and are considered as obligate parasites. Fungi by colonizing, multiply and survive on various habitats including micro ecological niches. Fungi also live in extreme habitats such as in deserts, thermal springs, hot springs, colder habitats and others.

Fungi occur in all aquatic, semi aquatic and arid habitats. They play very important role in recycling inorganic food substances. Their role in food chain formation is very significant. The fungi start colonizing temporarily or permanently on substrates of diversified nature including moist, submerged, wet and dried substrates. The process of degradation of organic matter by fungi is irreversible process and help in food chain to be used by flora and fauna living in water and on land.

Andhra Pradesh, ranks fifth in terms of area (2,76,700 sq.km.) in India and lies in between $12^0 40'$ and $19^0 50'$ North latitude and $84^0 40'$ East longitude. It is divided into three geographical regions viz., Coastal Andhra, Rayalaseema and Telangana for convenience. It has the longest coast line of around 1000 km., and abounds in two large rivers, the Godavari and the Krishna along with their several tributaries. The state is divided into 23 administrative districts. The climate of Andhra Pradesh in most part of the year is hot, mercury often touching 47^0 celsius. The state shows an average rainfall between 900mm-1100mm. It receives rainfall mostly from South-West monsoon in between June-September besides some amount of rainfall due to North-East monsoon between September-November. The main geological formations of the state are Deccan trap (Eastern ghats to Telangana), Gondwana formation (Godavari Valley) and Dharwar/ Peninsular granites. Mainly six types of soils manifest in the state viz., alluvial, black cotton, red laterite, arenacious and sandy loam soils.

Andhra Pradesh stands fourth in the country in terms of forest area having 63,777 sq.km. Region-wise, Coastal Andhra has 19,746 sq.km. (30.96%), Rayalaseema 14,951 sq.km. (23.45%) and Telangana 29,073 sq.km. (45.59%). The ranges of Forest of A.P. State are given below.

1) Forests of Eastern Ghats extending from Srikakulam to West Godavari through Visakhapatnam and East Godavari Districts.
2) Forests of Nallamalai hills found mostly in Kurnool District and a few pockets in Prakasam and Mahaboobnagar District.
3) Forest of Seshachalam hill range found mainly in Cuddapah and Chittoor District.
4) Forests of Telangana belt present in Adilabad, Warangal, Karimnagar, Khammam etc.

There are five types of forests occurring in the state.

They are :

1) Tropical dry deciduous forests (28,388 lakhs sq.km).
2) Tropical moist deciduous forests (16,100 lakhs sq.km.)
3) Dry tropical thorn forests (16,110 lakhs sq.km).
4) Littoral and dry evergreen forests (2,856 lakhs sq.km.).
5) Mangrove type of forests (3.45 lakhs sq.km.)

Fungi play an important role in human welfare. Fungi are also of biotechnological importance and biotechnological products include bakers yeast, antibiotics, organic acids, food processing, edible fungi, vitamins, growth promoters, fungal projects such as mycoremediation, biofertilizer production, waste water treatment, recycling of organic wastes and others.

Extensive survey of fungi were made by different researchers from all over the country and such fungi have been indexed in volumes named as Fungi of India. From Andhra Pradesh different workers have made extensive studies on the floristics, taxonomy and ecology of fungi from diversified habitats including the survey of fungi around aquatic bodies.

However, no attempt was made by them to compile and describe such fungi in the form of a manual or monograph but for the contribution of Rama Rao and Manoharachary's Soil Fungi from A.P. (1991). Therefore, an attempt is made to compile, index and describe the fungi around aquatic bodies located in different parts of Andhra Pradesh.

II Materials & Methods

II Materials and Methods

The fungi colonizing various substrates in/and around aquatic bodies were collected from Kawal wild life sanctuary, Sivaram Sanctuary, Pakhal and Eturnagaram Sanctuary, Kinnerasani, Papikonda, Kolleru Lake, Pocharam Sanctuary, Manjira and other places in Andhra Pradesh. The litter samples which include petioles, bark, leaf, dried flowers, fruits, damaged roots etc., were collected. The conidial fungi and Ascomycetes were collected at random, screened and numbered as per IMI pattern following Ellis (1961), Booth (1971a & 1971b) and Hawksworth (1974). The specimens of conidial fungi were screened under Lobo binocular microscope under artificial light. Each specimen was given accession number on a herbarium packet label giving the details such as name of the taxon, host or substrate, locality, name of the collector, date of collection and the relevant literature reference. The packets were prepared in duplicates and one set was sent to Herbarum Cryptogamae Indie Orientale (HCIO), IARI, New Delhi to obtain national accession number for publication.

The Myxomycetes were collected in small boxes or empty match boxes giving a support of tissue paper in order to keep the fruit bodies intact. These were also subsequently preserved as per IMI pattern. Packets were given the accession numbers.

The macro-fungi were collected from the field besides taking the photographs as far as possible in the natural habitats. This was followed by the morpho-taxonomic and anatomic studies of fruit body, spore print and histo-chemical examination.

The microscopic slides were prepared for conidial fungi by picking up the necessary portion of hyphae, conidiophores, conidial mat and other relevant structures and were studied in the stereo binocular microscope. Temporary mounts were made in lactophenol and warmed on a heater to drive away the air bubbles. The slides were sealed with quick fix or DPX mountant. Each slide was labelled giving the accession number.

The conidial seccession and germination were observed in cavity slides using sugar solution. The plates were prepared for each taxon depicting colonies, mycelium, conidiophores, conidiogenous cells and the conidia.

Whenever, required the conidial fungi were isolated on PDA. The cultures were observed under microscope. All the microscopic slides prepared were studied and photographed with the help of GETNER microscope. The slides were preserved in slide boxes wrapping them with the polythene bags in order to prevent them from the dust and for packets naphthalene balls were used to avoid insects.

Single spore isolations were made for some fungi which could be cultured. They were maintained on agar slants at 4^0 C in refrigerator. Others are maintained as herbarium specimens.

III Description of the Taxa

III Description of the Taxa

1. ACREMONIUM Link

Mag.Ges.Naturf.Freunde.Berlin 3:15, 1809

Acremonium was created by Link (1809), with *Acremonium alternatum* Link as its lectotype. *Acremonium* is a moniliaceous fungus and its allied genera *Cephalosporium, Gliomastix, Verticillium* and others are described. Therefore, the generic concepts of form genus acceptable to some mycologists may not be acceptable to others. Some *Acremonium* species as understood by Gams (1971), were transferred to *Gliomastix* by Ellis (1976), and Dickinson (1968). In general *Acremonium* is a moniliaceous fungus sometimes, produces coloured (dematiaceous), colonies are effuse, cottony, delicate, and dust to brown coloured. Mycelium superficial, creeping or sometimes forming mycelial mat or rope. Conidiophores micronematous or semi-micronematous, erect, branched, thin-walled, smooth. Conidiogenous cells monophialidic, single, erect, long, cylindrical, producing, conidia single or in chains/clusters. Conidia variable in shape, oblong, ellipsoidal.

Acremonium has been studied from all parts of the world. The most important monographic work is of Gams (1971), who has taken sufficient pains in compiling the *Acremonium, Cephalosporium* and allied genera under one cover. Bilgrami et al. (1991), from India have reported 28 species from soil, plant material as myco-parasite, litter infecting fungus. *Acremonium sp.* cause sugar-cane wilt in Uttar Pradesh, India, causes citrus fruit rot, spoil rhizomes of ginger in other places. Gams has recognized and categorised more than 70 species in 3 major sections which include *Verticillium, Gliomastix, Monocillium, Cephalosporium* and several ascogenous fungi-like *Nectrea, Hypocrea* and others having conidial state of *Acremonium.* It was also reported by Sukapure and Thirumalachar (1963), as *Cephalaosporium.* Presently, two *Acremonium* species viz., *Acremonium chrysogenum* (Thirum. & Sukap.) Gams, and *A. apii* (Smith & Ramsey) Gams, are being described.

Acremonium chrysogenum (Thirum. & Sukap.) Gams, Artige. Schimmelpilze. 109-111, 1971. (Fig. 1)

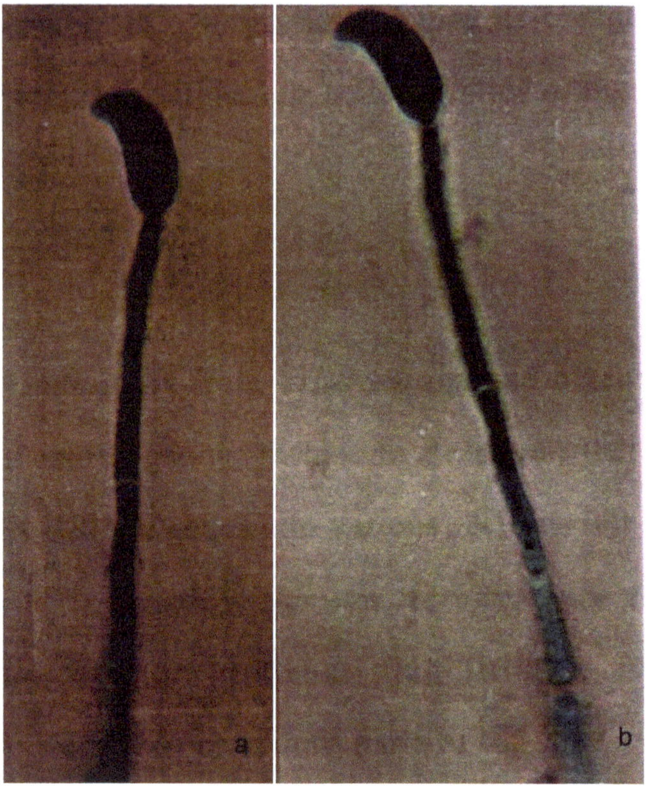

Fig.1. *Acremonium chrysogenum*
a-b, conidiophores with conidial attachment. (1000 x)

Colonies effuse, creeping, whitish or greyish in colour, hairy. Mycelium creeping, septate, branched, hyaline to sub-hyaline, thin walled forming mycelial rope. Conidiophores micronematous, mononematous, 10-15μm, 2-3μm broad, simple or unbranched. Conidiogenous cells phialidic, 15-35μm long, 2-3μm broad, cylindrical, monophilidic. Conidia solitary, 2-4μm long, 2-3μm broad, elongated or ellipsoidal.

On unidentified leaves collected from near by water bodies in Bhadrachalam, Yellandu forests of Khammam district, A.P., HCIO No: 42,996. NCBL No. 11, 87, 123. Nov. 2001. Coll. V.R. & G.S.K.

2. *Acremonium apii* (Smith & Ramsey) Gams, *Cephalosporium* Artige. Schimmelpilze. 136-138, 1971 (Fig.2)

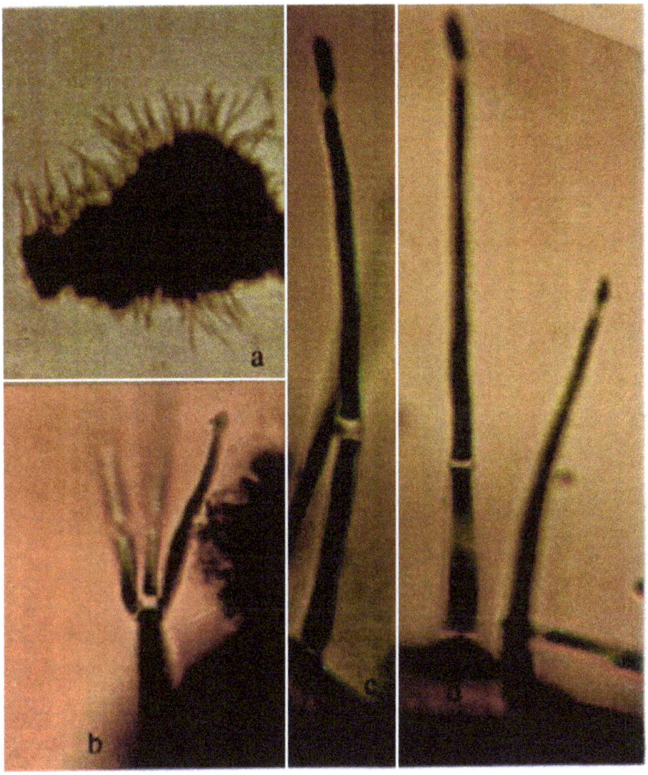

Fig.2. *Acremonium apii*
a.Colony (100 x), b.Conidiophores with conidial attachment (450 x)

Colonies effuse, whitish to brown, cottony. Hyphae creeping, hyaline to sub-hyaline, thin walled, 2-3μm broad. Conidiophores macronematous, mononematous, simple or branched, cylindrical, smooth, thin walled, 10-15μm long, 2-3μm broad. Conidiogenous cells phialidic, thin walled, cylindrical, colourless smooth, straight or curved, arising laterally on the mycelium. Conidia phialidic, enteroblastic, 0-1 septate, oblong, slightly bent in the upper region, 1-2μm broad, curved, smooth,10-15μm long and 2-4μm broad.

On unidentified leaves from around water bodies in Bhadrachalam, Ashwaraopet forests of Khammam district, A.P., HCIO No: 42,997. NCBL No. 66, 89. Nov. 2001. Coll. V.R. & G.S.K.

Acremonium is an important fungus and its significance is being studied throughout the world which is evident from the work of Quisenberry and Joost (1950) and Hawksworth et al. (1995). It is understood that *Acremonium* is alkalophilic and cellulolytic.

Key to the species of *Acremonium*

I. Conidiophore hyaline, branched/unbranched macronematous,solitary.Conidiogenous cells phialidic.Conidia aggregating at its tip, single celled.. *Acremonium*

1. Conidiophores unbranched, phialidic. Conidia ellipsoidal... *A. chrysogenum*
2. Conidiophores branched, phialides long. Conidia oblong.. *A. apii*

2. ACRODONTIUM de Hoog

Stud. Mycol. No.1: 23-27, 1972

Acrodontium was proposed by de Hoog (1972), with *A. crateriforme* (van Beyma) de Hoog, as its type species. It takes in forms producing velvety and powdery colonies. Mycelium septate, thin walled, branched, loose. Conidiophores arise orthoplagiotrophically from the creeping hyphae, sometimes thick walled, dark brown, verticillately or dichotomously branched. Conidiogenous cells orthotrophic or plagiotrophic in whorls, flask shaped or elongated, tapering, sympodial with denticulate rachis. Conidia one celled hyaline or pigmented.

de Hoog (1972), separated *Acrodontium* from *Tritirachium* and other genera on the basis of morphology of conidiogenous rachis. In *Acrodontium* it has alternate denticles. de Hoog divided the genus into various sections like *Acrodontium, Acrodontium* section and *Acrodontium griseum*. The *Acrodontium griseum* section produces plagiotrophic conidiophores. *Acrodontium* has been studied from various parts of the world isolated from soil, leaf material, food stuffs, dead spiders. It has been observed on the plant material as *A. griseum* (Fassatiova) de Hoog, which is being described here.

Acrodontium griseum (Fassatiova) de Hoog, Stud. Mycol. 1 . 36-37,1972. (Fig.3)

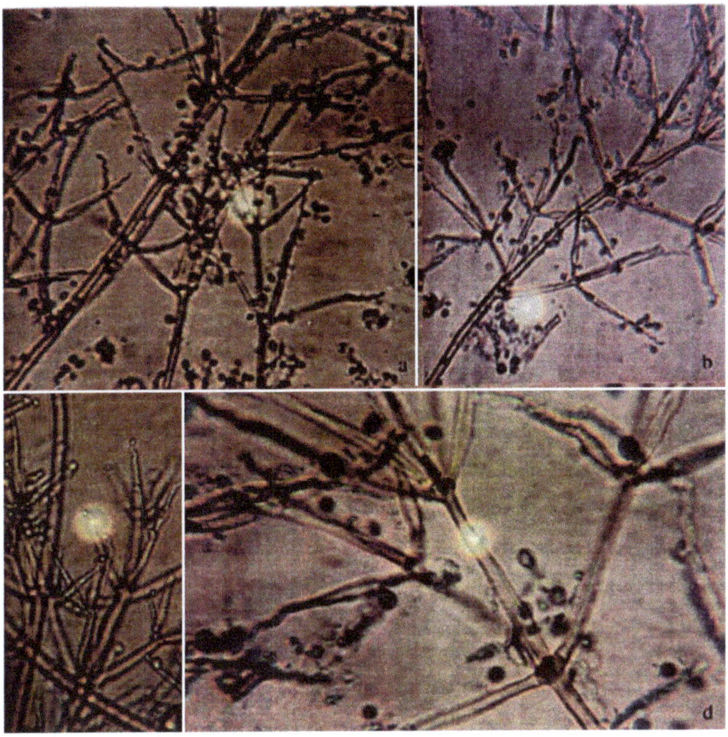

Fig. 3. *Acrodontium griseum* - a-d, Part of the conidiophores showing branching, conidiogenous cells with denticles and conidia (a-c 450 x, d 1000 x)

Colonies effuse, powdery, velvety. Mycelium superficial or forming small sporodochium-like cushions, 2-3μm broad, 5-15μm across. Conidiophores macronematous, mononematous, erect, simple, orthotrophic branched, at head region 50-300 μm long, 3-4.5 μm broad. Conidiogenous cells on the branches flask shaped, elongated, basal part swollen tapering towards tip, plagiotrophic in origin, pale brown, 5-15μm broad, geniculate, cicatrised, denticulate. Denticles at the intervals, hyaline to sub-hyaline, smooth walled. Conidia globose to sub-globose, guttuliform, fusiform with apiciulate base, 1.5-4.5μm long, 1.5-3μm broad, thin walled.

On decaying plant leaves from water bodies in Adilabad, Srikakulam, Khammam districts of A.P., NCBL No. 302, 214. Oct. 2001. Coll. V.R.&G.S.K.

Acrodontium griseum is the first record to India whereas *A. albaeo* was reported by Kushwaha and Agarwal, from soil in Sagar Madhya Pradesh.

3. ACROGENOSPORA Ellis

Dema. Hypho. 114-115, 1971

Ellis (1971), created *Acrogenospora* while presenting a beautiful account of dematiaceous hyphomycetes, from International Mycological Institute, Kew, UK, in a series of papers with *A. sphaerocephala* (Berk. & Br.) Ellis, as its type species. It is characterised by effuse colonies. Mycelium superficial or immersed. Conidiophores macronematous, mononematous, cylindrical, erect, simple. Conidiogenous cells monoblastic, percurrent, cylindrical. Conidia acrogenous, spherical or sub-spherical, obovoid or ellipsoidal, truncate at the base, black, shiny, smooth, zero - septate. This fungus has evoked a debate in the recent past because of its external similarities with *Monotosporella* Hughes.

Acrogenospora was originally collected on dead decaying wood in Europe and Australia. Matsushima (1975), studied *Monotospora sphaerocephala* Berk. & Br.which was included by Saccardo (1886), and Lindau (1907), and Hughes (1953). The specimens were collected in Europe, Canada and Japan on rotten wood. It has macronematous conidiophores, with percurrent simple and spherical conidia. Hughes (1958), created *Monotosporella* which resembled *Acrogenospora* superficially. Mason (1941), already studied the type material of *Monotospora sphaerocephala* Berk. & Br., and confirmed that it was anamorphic form of *Farlowiella.* Thus according to Hawksworth et al. (1995*), Acrogenospora* and *Monotospora* are identical however, clubbing *Monotosporella* with *Brachysporiella* is not acceptable to the authors, (sensu Hughes 1978), and it should be dealt separately for the obvious reasons. In the light of observations made by Vasant Rao and de Hoog (1986), Carmichael et al. (1980), and Hawksworth et al. (1995), there is a confusion in these genera and needs a serious consideration. Ellis (1971 and 1976), recognised *A. sphaerocephala* (Berk. & Br.) Ellis, *A. setiformis* (Walir.) Ellis, and *Farlowiella australis* Dennis state. Hughes (1978), while studying the New Zealand fungi redescribed *Acrogenospora sphaerocephala, Acrogenospora* state of *Farlowiella* and created two more species viz., *A. gigantospora* Hughes and *A. novaezealandiae* Hughes.

Acrogenospora sphaerocephala (Berk. & Br.) Ellis, Dema. Hypho. CMI Kew, UK. Pp. 114, 1971. (Fig. 4)

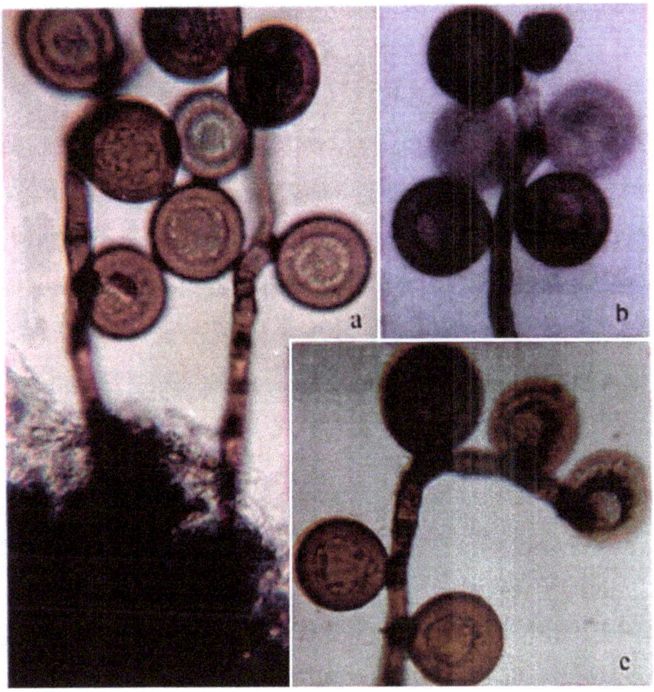

Fig. 4. *Acrogenospora sphaerocephala* - a-c, showing colony, conidiophores with percurrating conidiogenous cells, attached conidia (450 x)

Colonies effuse, black to brown, hairy. Hyphae superficial or immersed, 2-3μm wide and branched. Conidiophores macronematous, mononematous, erect, simple, long, 7-10μm thick at base, 5-7μm broad at apex. Conidiogenous cells terminal, integrated, percurrent, cylindrical,100-150μm long with 15-30μm broad, 10-30μm tall, truncate at base, 5-7μm broad at base. Conidia spherical or globose, smooth, pale to dark brown, 5-10μm in diam., still attached to conidiogenous cells.Conidial secession schizolytic, conidial germination from the base.

On unidentified, decaying wood and bark near water bodies from Bhadrachalam, Kottagudem forests of Khammam district, A.P., HCIO No:42,998. NCBL No. 13, 187. Nov. 2001. Coll. V.R.& G.S.K.

Acrogenospora Ellis, is a new record to Indian mycoflora. There is no published account on this genus from India. In literature as stated above there is a controversy about the systematic position of *Acrogenospora* Ellis, *Monotosporella* Hughes, and *Brachysporiella* Batista.

4. AGARICUS Linnaeus

ex Fries, Syst. Mycol. 1: 281, 1821

Agaricus was proposed by Linnaeus in 1753, with *A. campestris* Linn., as its type species. Fries (1821), accepted this genus, including his fungus *Psalliota campestris* Fr. as a synonym. *Agaricus* belongs to Agaricaceae and there are more than 200 species in this genus . It is recognised by its primary uninucleate mycelium which is confined to soil and short lived. The secondary mycelium is septate, binucleate branched, incorporated to form the basidiocarp. The young basidiocarps arise as small buttons which are covered by universal veil. The partial veil arises later and covers the stipe, increases in length. The partial veil arises later and covers the stipe and pileus. When the gills are fully developed it breaks up and remains as annulus on the stipe. The head of the pileus circular umbrella like white or pinkish carrying partial universal veil. The gills are free and are attached to the gill cleft. They contain trama, sub-hymenium and hymenium. The trama consists of loose hyphal strands with or without skeletal cells. Sub-hymenium is made up compact hyphal strands and skeletal cells. The hymenium is a fertile layer producing basidia and cystidia. The basidia are club shaped smooth and produce sterigmata and 2-4 basidiospores. The cystidia or paraphyses are sterile basidia like structures.

The genus *Agaricus* contains more than 200 species. It is cosmopolitan in distribution. The most common ones are *A. campestris, A. xanthodermus,* and *A. bisporus, A. campestris* is field mushroom and *A. bisporus* is cultivated mushroom. *A. xanthodermus* is poisonous. Bilgrami et al. (1991), have indexed 26 species from India. Chang et al. (eds. 1993), have discussed at length the cultivation, biology, ecology ones taxonomy and mushroom products of edible.

Agaricus campestris Linnaeus ex Fries, Syst. Mycol.1: 281,1821. (Fig. 5)

Fig. 5. *Agaricus campestris* - a-b, Basidiocarps. c-d, basidiocarp with button stage. e, button stage with universal veil. f, V.S. of basidiocarp. g, L.S. of gill. h-i, showing cystidia, basidia, basidiospores (g 450 x, h-i 1000 x)

Agaricus campestris is agaricoid in habit. Its plieus is 4.5-5 *cm* in fresh white when young becoming yellowish at center, margin entire, extended beyond lamellae, slightly in-curved or plane, appendiculate with marginal veil, context 3-5.4 *mm* thick, white, fleshy, non-confluent with fleshy stipe, non-hygrophanous, cuticle continuous in button stage, splitting into up-lifted scales on expanding, surface moist, but not viscoid, scaly with pale luteous scales becoming smooth and silky floculose with age apicuties thin walled hyphae 4-7μm broad parallel repent tramal hyphae thin walled hyaline 3-7μm broad septate. Lamellae free with unequal sects of four lamellulae delicate pink to chocolate brown, fleshy crowded 3-9 *mm* broad ventricose margin entire, trama homoisomerous regular thin walled 3-5μm broad, surface euhymenial, sub-hymenium celluar 10-15μm thick. Pleurocystidia absent. Cystidia when present basidia like, 22-25μm long, 7-10μm broad, clavate hyaline, thin walled, basidia 15-25μm long, 7-10μm broad, tetrasporic clavate, hyaline, thin walled. Sterigmata 2-3μm long, basidiophores 5-8μm long, 3-5μm broad, ellipsoidal, porate with distinct germ pore, apicule lateral, coccobrown, thin walled, smooth or thick walled.Stipe 36 *cm* long, 0.5-1.5 *cm* broad, cylindrical, slightly bulbous or atennulate base white or sordid brown, silky above ring or annulus glabrescent below or attnulate base superior white membrane context stuff becoming hollow hyphae, 5-10*mm* broad smooth and glabrous. The basidiocarp very delicate, white, destroyed by bacteria and ants into completely degraded small liquid with bad smell 10-12 hours after the collection.

From around water bodies in Bhadrachalam, Yellandu forests of Khammam, Vishakapatnam, Srikakulam districts of A.P. NCBL No. 215, 307, 310, 311. Oct./Nov. 2001. Coll. V.R. & G.S.K.

The present specimen belongs to *Agaricus* on account of fertile gill edge total absence of chilocystidia and lamella pink when young. It is in conformity, with *A. campestris* on account of simple annulus fungal odour white pileus and spores porate. The edibility tests were not conducted.

5. ARTHROBOTRYS Corda

Pracht. flora, P.43, 1839

Arthrobotrys was introduced by Corda in 1839, with *A. superba* Corda, as its type species. *Arthrobotrys* produces hyaline or pink coloured effuse colonies. Mycelium very delicate, thin walled, septate forming ring like traps. Sometimes conidiophores macronematous, solitary or in clusters, simple or branched. Conidiogenous cells terminal or intercalary, polyblastic, denticulate. Conidia holoblastic, hyaline, 2-3 celled on broad denticles. Due to renewed growth of the conidiophore conidial heads are formed which become intercalary.

Arthrobotrys has been studied by many people in several parts. There are 47 species recognised in this genus and many genera which look similar to *Arthrobotrys* like *Dactylella, Candelabrella, Genecularia* etc., as mentioned are included in this genus. *Arthrobotrys* occurs in the soil. It has been established that it is a predaceous fungus killing nematodes, rotifers, amoeba in water and soil. It has been reported from Canada, U.S.A., England,

Europe, Java, Australia and other countries. It has also been reported from India under the name *Arthrobotrys* and others. One of such species Subramanian (1971),very rightly said the fungus described by Bhattacharya and Baruah (1953), from India was neither *Pyricularia aquatica*, nor *Dactylaria* and suggested for its reclassification. *Arthrobotrys conoides* and *A.foliicola* are being described here.

1. *Arthrobotrys conoides* Drechsler, Mycologia, 29 : 447-552, 1937.

(Fig. 6)

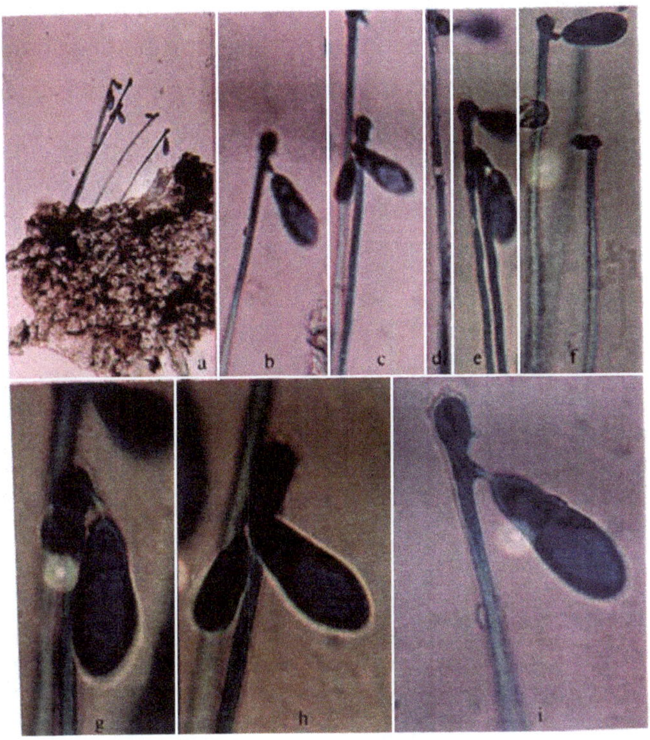

Fig. 6. *Arthrobotrys conoides* - a, colony (40 x). b-i, conidiophores and conidiogenous cells with denticles and conidia. (b-f 450 x, g-i 1000 x)

Colonies effuse, very light, cottony, whitish or pinkish in colour. Mycelium superficial or submerged in the substratum, thin walled, septate, 2-3μm broad forming ring like traps or so. Conidiophores erect, simple, macronematous, mononematous, cylindrical, septate, smooth, delicate, thin walled, 150-450μm long, 2-3μm broad. Conidiogenous cells terminal, polyblastic, denticulate, swollen, 3-7μm broad, terminal or intercalary, protuberance or denticles

2-5μm long, 1-1.5μm broad. Conidia holoblastic, pyriform to clavate, truncate at base, upper cell longer and larger than basal one, 15-30μm long, 5-10μm broad, smooth, colourless or pink. Conidial secession rhexolytic, conidial germination from the basal cell.

On dead moist leaves from water bodies in Medak, Adilabad, Karimnagar districts of A.P., HCIO No : 42, 999. NCBL No. 320, 210. Oct./Nov. 2001. Coll. V.R. & G.S.K.

In one of the specimens collected from near by other water bodies, the conidiophore, conidiogenous cells and conidia are similar to *Arthrobotrys conoides.* The terminal cell produces short protruberence whether it is a process of germination or germ tube could not be established, tentatively, it is treated under the same genus as second collection.

2. *Arthrobotrys foliicola* Matsushima, Icones Fungorum, 10, 1975.

(Fig. 7)

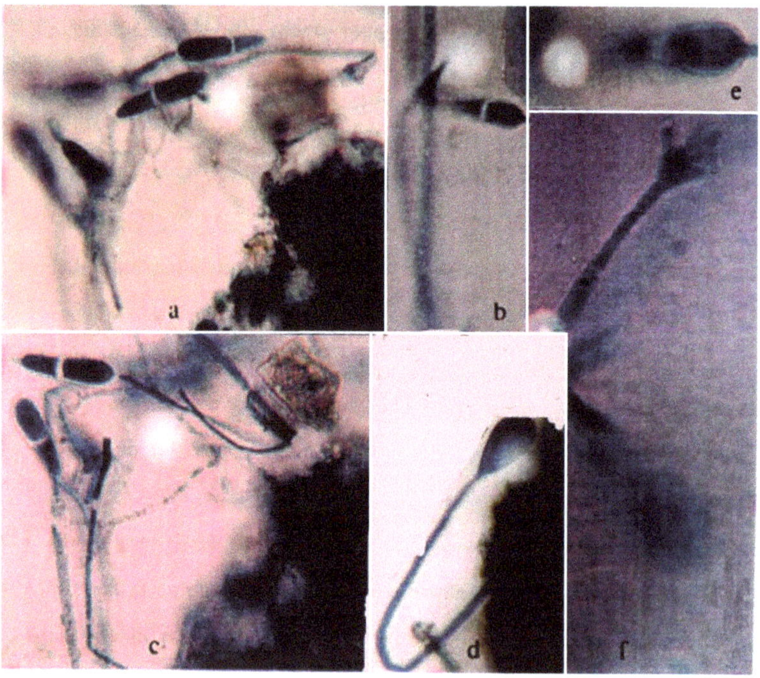

Fig. 7. *Arthrobotrys foliicola* - a-f, colony, conidiogenous cells, conidiophores with conidia. (a-d 450 x, e-f 1000 x)

Colonies on decaying leaves effuse, minute, cottony. Mycelium immersed, septate, thin walled, smooth, 2-3μm broad. Conidiophores macronematous, mononematous, solitary or in fascicles, erect sample, septate, smooth, 100-200μm long swollen 2-5μm broad. Conidiogenous cells polyblastic, denticulate, sympodial due to proliferation of the conidiophore nodes and internodes from every short conidiogenous region 3-7μm flat, 5-7μm long with distinct, 1-3μm long denticles upto 1.5 μm broad. Conidia holoblastic, obovoid, cylindrical, smooth, 1-3 septate, 10-15μm long, 3-4.5μm broad base and truncate, apex pointed or narrower.

Collected on moist leaves from near by water bodies in Medak, Adilabad, Nalgonda districts of A.P., HCIO No: 43,000. NCBL No. 209, 317. Sept./ Dec. 2001. Coll. V.R. & G.S.K.

Arthrobotrys is a predaceous fungus reported throughout the world. It is most studied fungus in Applied Mycology. It has been observed to kill the nematodes in the soil and other aquatic organisms. In India little attention has been paid for the applied aspect of such fungi.

Key to the species of *Arthrobotrys*

I. Conidiophores macronematous, mononematous. Conidiogenous cells with nodes polyblastic, denticulate. Conidia 2-celled, ellipsoidal/obovoid oval or clavate or pyriform....................................... *Arthrobotrys*

1. Conidia pyriform or clavate.. *A. conoides*
2. Conidia obovidal, pointed base...................................... *A. foliicola*

6. BAHUGADA Reddy and Vasant Rao,

Cur. Sci. 54 : 147-149. 1984

Reddy and Vasant Rao (1984), proposed *Bahugada* with *B. sundara* as type species. This fungus occurs on the barks of the fallen trees of *Borassus*. It is characterised by geniculate conidiophores bearing polyblastic, sympodial denticulate, conidiogenous cells, producing pyriform or obclavate muriform conidia.

Originally *Bahugada* was described from Orissa state (India). This monotypic genus sometimes confuses with *Berkleasmium* which produces stroma and uncommon conidiophores not found either in *Oncopodiella* Arnaud ex Rifai and *Monodictys* Hughes. The type species is being described here which forms a new record to Andhara Pradesh.

***Bahugada* sundara** Reddy & Vasant Rao, Curr.Sci. 54 : 147-149, 1984

(Fig. 8)

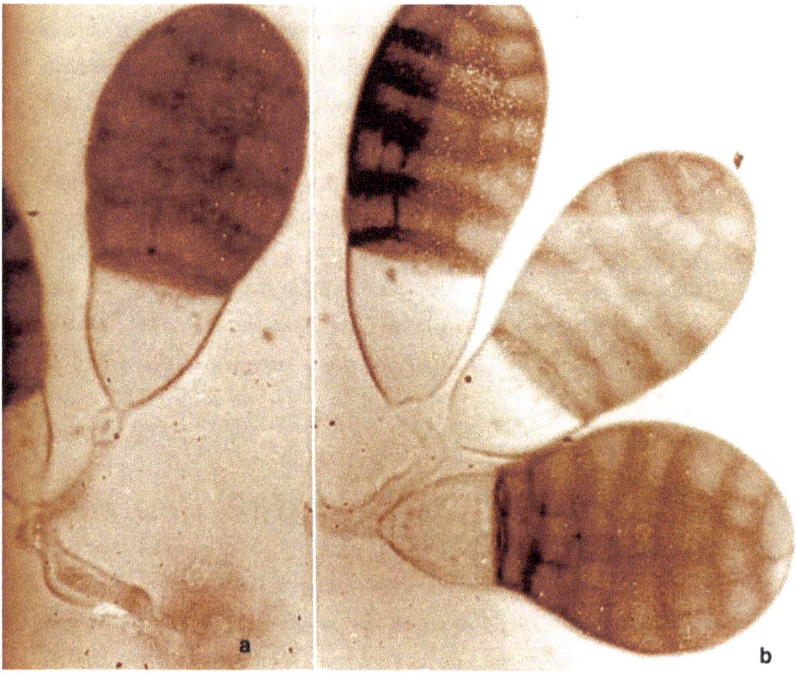

Fig. 8. *Bahugada sundara* - a-b, Conidophores with conidia (1000 x).

Colonies effuse or puntiform, dark to blackish brown. Mycelium immersed, composed of branched septate, smooth subhyaline to pale olivaceous brown, 2-3.5μm thick hyphae. Conidiophores arising singly, laterally and terminally from the hyphae, macronematous, mononematous, simple or very rarely branched, erect, geniculate, subhyaline becoming hyaline towards apex, smooth, septate, upto 60μm long,1-3.5μm thick. Conidiogenous cells polyblastic, integrated, terminal, becoming intercalary, sympodial, denticulate, denticles cylinderical to conical. Conidia holoblastic, solitary, dry, acropleurogenous, smooth, pyriform or obovate with a proximal cyathiform basal hyaline cell, 5-10μm long, 8-10μm broad. Basal cell is distinctly separated from a distal part in dark brown, pigmentation and longitudinally and transversely septate portion of the conidium. Conidia are 20-40μm long, 10-20μm broad at the broadest. No germination could be observed.

Collected on *Hardwikia binata* stem, *Borassus* bark & *Acacia arabica* bark Devarkonda, Nalgonda, Warangal, Karimnagar, A.P. NCBL No. 226, 308. Oct./Nov. 2002. Coll. V.R. & G.S.K.

7. BERKLEASMIUM Zobel

in Corda's Icones Fung. 6: 4, 1854

Berkely described *Sporodesmium concinnum* in 1845, (in Hooker's London J. Bot.4:309), which was later made *Berkleasmium coredeanum* by Zobel in 1854. Moore (1958), re-established the genus *Berkleasmium*,and later Moore (1959), prepared a monograph on *Berkleasmium* with 10 species including the *B.concinnum* (Berk) Moore, as its type species. *Berkleasmium* produces sporodochial colonies. Mycelium immersed, septate, branched, stroma, pseudoparenchymatous. Conidiophores macronematous, semi-macronoematous,erect, simple. Conidiogenous cells terminal, monoblastic. Conidia muriform, variable in shape, terminal. Hughes (1958), proposed *B. cordeanum* = *B. concinnum,* the present type species.

Berkleasmium is a conidial fungus and it has been distributed with 12 species in Europe, North America, India, Africa and South America. Goos (1969), and Moore (1958 and 1959), have presented a detailed account of the genus from various parts of the world. From A.P., Raghuveer Rao and Dev Rao (1964), have reported two species namely *B. papilatum* and *B.osmaniae* as new speices. Holubova-Jechova and Kirk (1984, 1985) reported *Berkleasmium* from South Africa. *B.pulcherum* Holubova-Jechova, was recollected and described by Vasant Rao and de Hoog (1986). This fungus has affinity with *Cryptocorymium* Fuckel. *Berkleasmium* produces sporodochial colonies, short macronematous, conidiophores, monoblastic, conidiogenous cells muriform or dictyospores, variable shaped conidia. A sporodochial fungus with muriform elongated conidia, but hyaline, sub-hyaline not black or opaque fungus similar to *Berkleasmium osmaniae,* was collected which is being described here.

Berkleasmium osmaniae Raghuveer Rao and Dev Rao. Mycopath, Mycol. Appl. 22:313-314,1964,1968. (Fig. 9)

Fig. 9. *Berkleasmium osmaniae* - a, Colony (100 x). b-d, Conidiophores with conidiogenous cells and conidia. (b-c 450 x, d 1000 x)

Colonies sporodochial, light whitish brown, velvety. Mycelium immersed, 2-3μm wide, thick walled incorporated in the formation of pseudoparanchymatous stroma. Stroma erumpent, thick walled, hyaline to sub-hyaline to light pale brown producing conidiophores. Conidiophores macronematous, simple, short, cylindrical, unbranched, septate, 3-5μm broad, 7-20μm long, hyaline to sub-hyaline, smooth. Conidiogenous cells monoblastic, terminal, holoblastic, integrated, cylindrical, one-celled. Conidia elongated, monoblastic, oval to clavate or cylindrical in shape, 20-70μm long, 3-5μm broad at base, 5-10μm broad in middle, 3-7μm broad at the apex, transversely and longitudinally septate, smooth, hyaline in the septal region, hyaline to sub-hyaline thin walled.Conidial secession rhexolytic germination could not be found.

On decaying plant debris near water bodies from Bhadrachalam, Yellandu forests of A.P., HCIO No. 43001. NCBL No. 97, 220. Nov./Dec. 2001. V.R.& G.S.K

Berkleasmium osmaniae described above is slightly different from the type in pigmentation only. In the type material conidia are blackish brown irregularly septate, oval to clavate and cylindrical shape. Recent trend in the taxonomy is based on the conidial ontogeny and not the colour and septation as considered earlier, the authors prefer to the conserve and place their taxon in the earlier known anamorphic form. It is also not known whether it is juvenille state of *B. osmaniae,* however, the collections made in successive seasons revealed the same characters.

8. CIRRENALIA Meyers & Moore

American J. Bot. 47 : 346, 1960

Cirrenalia was proposed by Meyers and Moore (1960), for a single species namely *C. macrocephala* (Kohlmeyer) Meyers and Moore, which is a synonym of *Helicoma macrocephala* Kohlmeyer. *Cirrenalia* was again emended and redescribed by Kohlmeyer (1966). Presently *Cirrenalia* is recognised by pulvinate, punctiform, effuse, reddish brown to black colonies. Mycelium superficial, septate, stroma weak or none. Conidiophores micro or semi micronematous, sometimes with inflated cells. Conidiogenous cells monoblastic, integrated, terminal.Conidia acrogenous, solitary, helicoid, septate, reddish brown to dark brown, smooth or rarely ornamented.

Cirrenalia Meyers and Moore, is marine and terrestrial conidial fungus degrading moist plant material. Sutton (1973), while studying the hyphomycetes from Manitoba and Saskatchewen in Canada proposed *C. donnae* in recognition of Donnae Brahimiella for her enthusiastic help in collection of micro fungi. He also described and took into consideration the marine species of Kohlmeyer namely *C.pygmea* Kohlmeyer (1966), *C. pseudomacrocephala* Kohlmeyer, and *C. tropicalis* Kohlmeyer (1968). These 4 species are moniliaceous producing conidia from rather poorly developed repent ramifying hyphae. The conidiophores, conidiogenous cells and conidia are not aggregated into punctiform sporodochia which are characteristic feature of *C. donnae.* Vasant Rao and Reddy (1978), proposed *C. indica* as a new species from India. Goos (1985), reviewed the so far described species and proposed a key to the recognised species. Thus *Cirrenalia* is known to India, America, Canada and other places.

Cirrenalia indica Vasant Rao and Reddy, Indian J. Mycol. Res. 306-308, 1978. (Fig. 10)

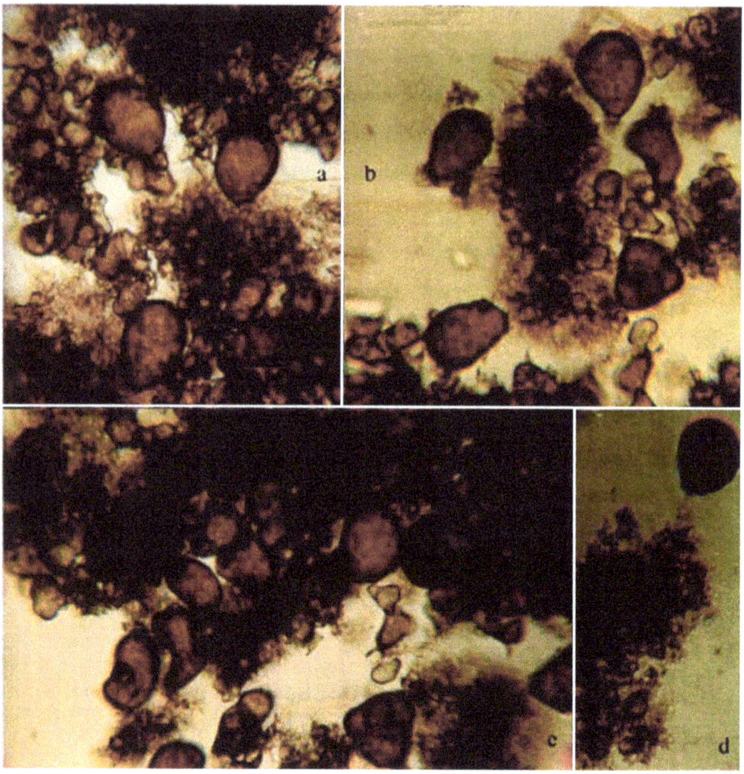

Fig. 10. *Cirrenalia indica* - a-d, Colonies with conidophores, conidiogenous cells and conidia (450 x).

Colonies effuse, punctiform, granular. Mycelium superficial, septate, scantly developed, smooth, 2-7μm broad, irregularly swollen or tubular thin or thick walled.Conidiophores micro or semi-micronematous, very small 2-5μm broad, sub-hyaline, pink or reddish brown. Conidiogenous cells terminal, integrated, discrete or indiscrete, monoblastic, 2-5μm long, 2-4μm broad, cylindrical or small tubuler smooth. Conidia acrogenous, monoblastic, holoblastic, 2-3 celled, first 2 cells very small, last cell profusely inflated, 5-15μm long, 2-3μm broad at base, 5-10μm broad at the broadest, rounded apex, reddish brown, thick walled, smooth, curved. Conidial secession rhexolytic, conidial germination from the base.

On unidentified leaves from water bodies in Medak, Adilabad, Yellandu, Kottagudem forests of A.P., HCIO No:43,003. NCBL No. 27, 135. Nov. 2001. V.R. & G.S.K.

Cirrenalia is being collected for the second time in India. Earlier Vasant Rao and Reddy (loc. cit.), reported it from decaying wood collected in zoo park, Hyderabad. The authors examined type material VMRL. No: 297, and compared the species with the Khammam and other material which differs in minor characters not worthy of specific or varietal status. Such as the cell of the *Cirrenalia* described by Vasant Rao & Reddy (1978), are slightly variable in shape.

9. COMATRICHA Preuss 1851

Linnaea 24: 99-153 emend., Nann - Brem.k.Ned.Acad.Wet.Proc. C.70, 206

Comatricha was proposed by Preuss in 1851, with *Stemonitis obtusata* Fries, as its type species. *Comatricha* produces fructification stipitate, sporangia, globose to cylindrical peridium fugaceous, columella halfway reaching nearly the apex of the sporangium, capitilium lacks and dense arising from the entire columella branched some what at right angles anastomosed and intricately looped with few anastomoses with few are many threads.

Comatricha has been recognized in India by Lakhanpal and Mukerji (1981), with 13 species.

Comatricha aequalis Peck, Ann. Rep. N. Y. State. Mus. 31: 42, 1879. (Fig. 11)

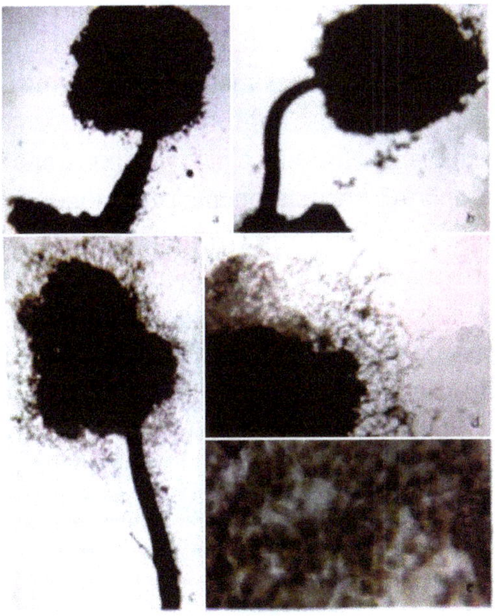

Fig. 11. *Comatricha aequalis* - a-e, Fruit body, capitulae, spores. (a-c 100 x, d 450 x, e 1000 x).

Colonies fructifications, sporangiate, stipitate, 4-5*mm* in length sporangia, densely clustered erect or flexuous and accurate with acuminate apex, dark purpulish brown to almost black. 2-3 *mm* in length peridium evanescent, collumellae stout same coloured with stipe branching into the capitilial net above or dendroyed towards apex. Capitilium form the entire collumellae primary branches prominent or otherwise stout and expanded at base branched and anastomosed to form an intricate internal net surface, net delicate and fragmentary small meshed free ends, numerous, pale or hyaline, spore mass purplish brown. Spores globose, 5-7μm in diam., warted stipe black, dull or slightly shiny,1.5-2.5 *mm* long, hypothallus dull brownish black and confluent.

From a water body on litter in Bhadrachalam forest of A.P., HCIO No: 43,004. NCBL No. 50, 289. Aug. 2002. V.R. & G.S.K.

Comatricha aequalis was originally collected in Cat-skill mounts, New York,U.S.A. It was collected on decaying coniferous wood. It is distributed in Europe.N.America, Jamaica and in India.

10. CYLINDROCLADIUM Morgan

Bot. Gaz. 17: 190-192, 1892

Cylindrocladium was created by Morgan(1892),with *C.scoparium* Morgan,as its type species. It was initially collected as saprophyte on rotting leaves, but later it was found to be parasitic on economically important crop plants. It is characterised by effuse light brown or whititsh colonies. Mycelium superficial, septate,branched forming chlamydospores and sclerotia in culture and nature. Conidiophores are macronematous, mononematous,erect,in the inner part. Repeatedly forming solitary branches and ultimately densely, compacted, pencillate groups of conidiogenous cells which are phialidic, doliiform, the main axis mostly long unbranched thread, ending in sterile swollen vesicle. The conidia are phialidic, aggregated in parallel bundles, cylindrical with symmetrically rounded at either ends or almost truncated ends straight or curved, 1- many septate, smooth and colourless or white.

Cylindrocladium is a polyphagus plant parasite. It occurs on *Picea, Prunus. Citrus,* neem and other plants. There is a considerable phytopathological literature on the pathogenticity, control of *Cylindrocladium* species. Boedijn and Reitsma (1950), have written a monograph on this genus. After studying all materials collected from different parts of the world they accepted only seven species. Rossman (1979), expanded the genus markedly by about 15 species of which several are connected with *Calonectria,* (Ascogenous teleomarph). Peerally (1974), reported many species of *Cylindrocladium* from Mauritius and other countries. Bilgrami et al. (1991), listed ten species reported from India on various crop plants like Cabbage, *Brassica, Wisteria, Cocus. Eucalyptus, Spinach, Eugenia, Madhuca, Ficus, Anacardium. Medicago, Dalbergia, Eicchornia, Thea* etc.

Cylindrocladium scoparium Morgan, Bot. Gaz. 17 : 190-1 92, 1892.
(Fig. 12)

Fig. 12. *Cyilindrocladium scoparium* - a-d, Conidiophores, conidiogenous cells and conidia (a-b 100 x, c-e 450 x, f 1000 x).

Colonies effuse, circular, brown with yellow halo, discrete. Mycelium superficial, immersed, septate, branched, sometimes produces chlamydospores and sclerotia in natural substrate or culture, smooth, 3-4μm broad. Conidiophores macronematous, mononematous, erect, simple, main axis branched in the middle region containing solitary branches

150-250μm long, 2-3μm broad, septate, bearing 2-4 branches. Conidiogenous cells phialidic arranged in pencillate manner, each phialide cylindrical, smooth, 3-9μm long, 2-5μm broad, hyaline to sub-hyaline. Conidia phialidic elongated, smooth, 1-septate, 6-15μm still long, 3-6μm broad, hyaline to whitish. No secession of conidia and germination was observed.

On decaying leaves Medak, Adilabad, Khammam, Srikakulam districts of A.P., HCIO No:43005. NCBL No. 167, 282. Oct. 2001. Coll. V.R.&G.S.K.

This *Cylindrocladium* is a pathogenic and parasitic fungus to many economically important plants. It causes leaf spot diseases and is responsible for substantial economic loss to ornamental and economically important plants. From India 10 species have been described which are named after their hosts. The host range of *Cylindrocladium* is Cabbage, *Brassica, Cocus, Eucalyptus, Euginea, Ficus, Dalbergia, Anacardium* and *Arachis hypogea* etc. *Cylindrocladium* infects the roots, leaves and stems. It's control is possible by use of common fungicides.

11. DALDINIA Cesati and De Notaris

Com. Soc. Brit. Ital. 1: 177-240, 1863

Cesati and De Notaris (1863), created *Daldinia* with *D. concentrica* (Bott. ex Fri) Cesati and De Notaris, as its type species. It is characterised by superficial, hemispherical to globose or clavate, sessile to substipitate, stromata with conspicuously concentratically multizonate, monostrictous rarely polystrictous, crowded, cylindircal or clavate to broad ellipsoid or sometimes globose, completely immersed perithecia and uniseriate smooth in equilateral, ellipsoidal, brown one-celled ascospores.

Daldinia belongs to Ascomycotina, Xylariaceae. It consists of 13 species distributed throughout the world. Cosmopolitan in distribution popularly known as calicowood or cramp balls. In India there are 6 species distributed in North, South, East and West regions. Thind and Dargan (1978), have given an excellent account for Indian species. They recognised and dealt with six Indian species giving a key for their identification. During the studies on macro and micro fungi of Khammam the type species i.e. *D. concentrica* was collected which is being described here.

Daldinia concentrica (Bolt. ex Fr.) Ces. & De Not., var. *escholzii* (Ethrenb.) Starb. Bihang K. Svenska Vet. Acad. Handl. Abd. III 27:5, 1901. (Fig. 13)

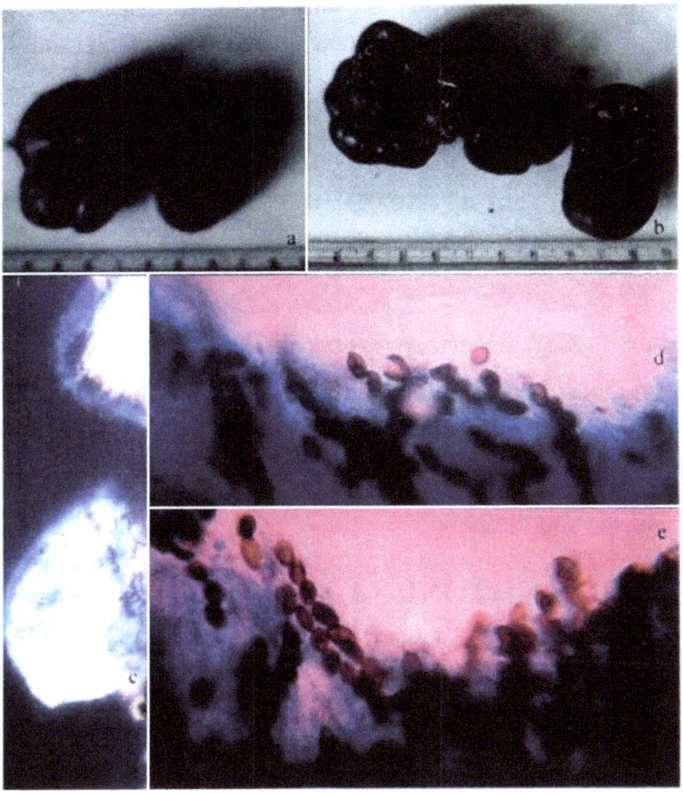

Fig. 13. *Daldinia concentrica* - a-b, Ascomata. c-e, V.S. of ascocarp and ascospores. (c 450 x, d-e 1000 x)

Ascocarp or ascomata, telemorphic, stromatic, hemispherical, globose or pulvinate, 1-5*cm* long, 1-3*cm* broad, 0.9-2.5*cm* high, scattered or aggregated single or confluent and deformed by mutual pressure, flattend, broad at the base, mostly sessile rarely stalked or stipitate, surface smooth, reddish brown to dark brown and finally black and shining. Mycelium immerrsed in the host tissue and incorporated in the formation of fruit body thick walled prosenchymatous, ectostroma not observed, endostroma massive, persistant, corky with well developed concentric zones, light and dark zones alternating, light zones greyish brown almost double than the dark zones, dark zones brown narrow, perithecia upto 1*mm* into 300 *mm* perithecial wall thick, textura angularis, cells very large thick wall not differentiated into zones, immersed ovate monostrictous closely placed ostioles punctate, ambilicate, sometimes raised above the stromata. Asci cylindrical, long stalked apex, obtuse, ascal plug weak, discoid, intact, asci could not be picked up. Asco-spores inequilateral, ellipsoid, dark black brown with germ slit, no paraphyses were observed, no anamorphic state was observed.

Material on decaying wood logs from Bhadrachalam forest of Khammam district, A.P., HCIO No. 43,049. NCBL No. 32, 85, 165. Nov./Dec. 2001. Coll. V.R. & G.S.K.

Ehrenberg (1820), described this taxon as "Sphaeria", but Child (1932), treated it as *Daldinia* species, Dennis (1963), considered this as a tropical fungus from India. It was earlier reported from Bihar, M.P., U.P. and Punjab (cf. Bilgrami et al. 1991).

12. DEIGHTONIELLA Hughes emend D.Shaw

Papua New Guinea agric. J. 11(3): 77, 1959

Hughes (1952), created *Deightoniella* with *D. africana* Hughes, as the type species and suggested a new combination. viz., *D.arundinacea* (Corda) Hughes. The fungus produces proliferating nodulose simple or branched conidiophores bearing 1-many septate or pseudoseptate conidia, variable in shape.

In 1957 Ellis redescribed the type species along with five more species in *Deightoniella* as *D. indica* Agarwal and Hasija (1961) from India, as *D. jabalpuriensis,* Satya and Rajya Lakshmi (1965), as *D. bhopalensis, D. torulosa* (Sydow) Ellis was recorded by Kapoor & Tandon (1968), which was also reported in air by Sree Ramulu (1962), and Sree Ramulu and Ramalingam (1964). Subramanian (1971) however, raised *Deightoniella indica* Subram., to a new genus viz., *Bahusagandha indica* (Subram), as the type . Ellis (1976) described a new species, *D. ichnanthi* Ellis from Guyana. In the present investigation *D. torulosa* is being described from the water bodies in Miryalaguda and other places.

Deightoniella torulosa (Syd.) Ellis, M.B. 1957 Mycol. Pap; 66 : 7-10.
(Fig. 14)

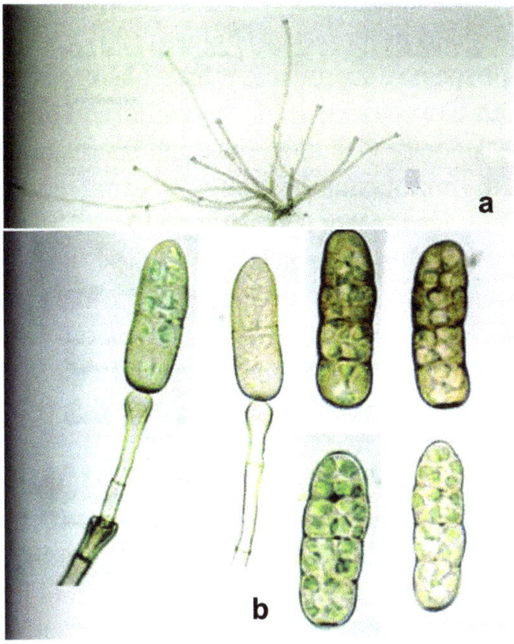

Fig. 14. *Deightoniella toruloa* - a, Colony (100 x).
b, Conidiophores with conidia (450 x). c, Conidia (1000 x)

Colonies of this fungus effuse, brown to black. Mycelium immersed, consisting of hyaline to pale brown, smooth, septate, 2-5.1*µm* thick hyphae. Conidiophores macronematous, mononematous, torsive, unbranched, brown, smooth , thick walled, 3-1, 96*µm* long, 8.1-10.2*µm* wide at base. Conidiogenous cells monoblastic integrated, terminal, percurrent, doliiform or subspherical, 2-3 percurrations, 13.2-16.3*µm* in diam., after falling of first conidium the upper part of the conidiogenous cells collapse and the conidiophore proliferates straight through an old scar. Conidia solitary, acrogenous, simple, obpyriform or obclavate, straight or slightly curved, pale to mid brown, smooth, with 3-5 transverse pseudoseptate, 35-65*µm* long, 13.2-20.3*µm* wide at the broadest.

Collected on *Albizzia amara* stem, near water bodies Nereducherla, Nalgonda, Warangal, Adilabad, A.P. NCBL No. 14, 56, 140. Nov. 2001, Coll. V.R. & G.S.K.

It is a very strange fungus differing from other described species but very close *D.torulosa*. It is a new record to mycoflora of Andhra Pradesh.

13. DENDRYPHIOPSIS Hughes

Canadian J. Bot. 31: 577-659, 1953

Dendryphiopsis was proposed by Hughes (1953), with *D. atra* (Corda) Hughes, as its type species. It produces effuse, black, pin head size colonies, black pins with fan-like conidial head. Mycelium creeping on the substrate, branched, septate, smooth. Conidiophores macronematous, mononematous, arborescent with numerous branches bearing conidiophore cells. Conidiogenous cells terminal or lateral, enteroblastic, tretic, elongated with prominent black scars with a pore in it. Sometimes, the conidiogenous cells arise below the conidial scar of the first conidium in a zig-zag manner and conidiophores head becomes profusely branched. Conidia phragmosporous, 3-4 septate, smooth, cylindrical. *D. arbesculum* (Berk. & Curt.) Hughes, as another species reported. Ellis (1971), reported this taxon on number of substrates from Common Wealth countries. Matsushima (1971), recorded the same from Japan and Solomon Islands. Mercado et al. (1997), rediscovered *Dendryphiopsis* from Cuba, with its perfect state *Microthalia* which was synonymous with *Amphisphaeria.* This taxon has been reported on *Carya*, *Drypetes, Euphorbia, Magnolia,* and other host plants.

Dendryphiopsis atra (Corda) Hughes, Canadian J. Bot. 31: 655, 1953. (Fig. 15)

Fig. 15. *Dendryphiopsis atra* - a, Colony (100 x). b-g, Conidiophores, conidiogenous cells with conidia (b-e 450 x, f-g 1000 x)

Colonies effuse, dispersed, black, pin head like with a fan like structure. Mycelium immersed in the substrate, hyphae, septate, branched, pale to dark brown, thick walled, 3-5μm broad. Conidiophores macronetmatous, mononematous, erect, straight or flexuous, branched in the apical region looking arboresent (tree like), 200-250μm long, 3-5μm broad, 100-120μm broad in the tree like region. The branches arise below the septa like the conidiogenous cells. Conidiogenous cells terminal or lateral or in primary or secondary or tertiary in order, 0-1 septate, elongated, having pore at terminal or lateral portion, 10-20μm in length, rounded at the apex, base lightly depressed. Conidia enteroblastic, tretic, 3-4 septate, elongated, septa band like, thick walled, 30-60μm long, 11-20μm broad, single on the conidiogenous cells. Conidial secession rhexolytic and germination through basal pore or through appical cells.

On decaying and unidentified plant debris collected from Adilabad, Medak, Karimnagar & Warangal, Bhadrachalam, Yellandu, Ashwaraopet forests of A.P., HCIO No: 43,006. NCBL No. 5, 67, 136. Oct. 2002. Coll. V.R. & G.S.K.

Dendryphiopsis atra is a common form genus in many tropical, subtropical and temperate countries. In India according to Bilgrami et al. (1991), it has been reported from Maharashtra and Kerala. There is slight variation in the conidial configuration and pattern of formation of the lateral conidiogenous cells. It reminds other genera like *Edmondmasonia, Kramabeeja, Brachysporiella* and other genera. However, they differ from *Dendryphiopsis* in their conidial ontogeny. There is no authentic work on its enzymes but it has been collected on number of occasions on rotting twigs, wood, and bark. It could be presumed that it may be lignicolous fungus with a capacity to degrade lignin, cellulose and pectin.

14. DICTYOARTHRINIUM Hughes

Mycol. Pap. 48: 30, 1952

Dictyoarthrinium was proposed by Hughes, with *D. quadratum* as its type species. It produces effuse granular black colonies. Hyphae septate, branched and anastomosing. Conidiophores basauxic, macronemetous, mononematous, erect or curved. Conidiogenous cells terminal or intercalary. Conidia terminal and lateral, four celled to many celled, cruciform or variable in shape.

Hughes (1952), proposed *Dictyoarthrinium* with *D. quadratum* and *D. africanum* collected on leaf litter from Africa. Damon (1953), observed that there existed another fungus agreeing with the description of the *Dictyoarthrinium,* described as *Tetracoccosporium sacchari.* Therefore,

Damon suggested *D. sacchari* (Steve. ex Johnson and Stevenson) Damon for consideration. Hence, instead of *D.quadratum, D.sacchari* was proposed as type species. *D.africanum* collected from Africa produces many celled conidia. *D.sacchari* was reported by Subramanian (1952) and Rao and Rao (1964), described this type species from India and also created new species like *D. lileputium* Rao and Rao which is nothing but the juvenile form of *D. sacchari* (vide von Arx, Pers. com.). *Dictyoarthrinium* has been isolated from *Borrasus, Bamboo and Musa* leaves. It has been also isolated by Matsushima (1971), from soil and reported it as a new species namely *Dictyoarthrinium raboulense* Matsushima. Kulkarni (1984), worked on *Dictyoarthrinium* and allied basauxic genera. His studies were based upon CAB (CMI) cultures and specimens. *Dictyoarthrinium* has been isolated from soil, plant debris and air. *Dictyoarthrinium* is a foliicolous and lignicolus fungus. It has been reported from India, Cuba, Africa, Japan and other countries.

Dictyoarthrinium sacchari (Steve. ex Johnson and Stevenson) Damon, Bull. Torry, Bot.Clb.164, 1953. (Fig. 16)

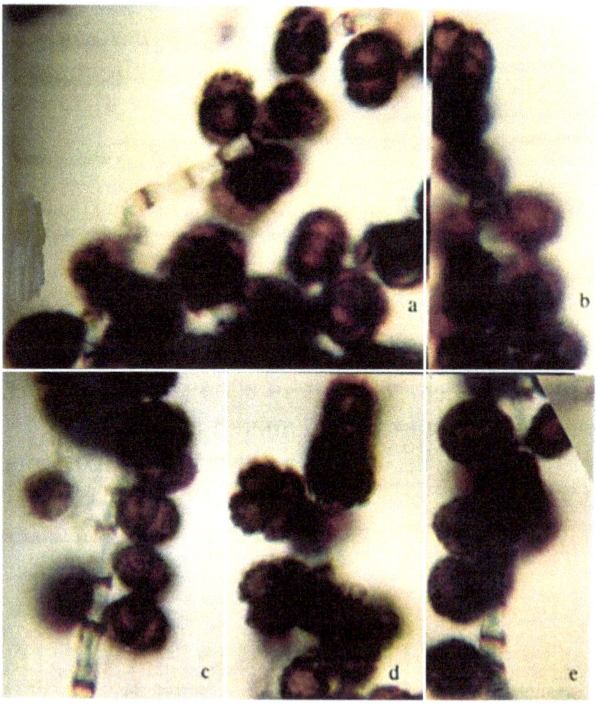

Fig. 16. *Dictyoarthrinium sacchari* - a-e, Conidiophores, conidiogenous cells and conidia (1000 x).

Colonies effuse, granular. Mycelium superficial, anastomosing, hyphae creeping on the substrate, septate, branched, 1.2-4.5μm broad, anastomosing. Conidiophores macronematous, mononematous, simple, septate, 100- 150μm in length, producing, conidia terminally and laterally. The first formed conidium is always found on the top of the conidiophore as a result of transversely breaking of conidiophore mother cell and elongating intercalarily, subsequently conidia are produced laterally. Lateral conidia arise as blown out ends. Conidiogenous cells terminal or intercalary in between the bands of the conidiophore, 5-10μm long, 2-3μm broad, smooth, hyaline to sub-hyaline. Conidia 4 celled terminal or lateral, with half of the conidiophore mother cell at the tip and base of first conidium which is holoblastic, another half of the conidiophore mother cell remains at the base of the conidiophore. The conidiophore lies in between the two halves of the conidiophore mother cell.

From leaf litter from all over A.P. Districts, HCIO No: 43,007. NCBL No. 218, 315. Nov. 2002. Coll. V.R. & G.S.K.

Cole (I974), Hughes (1953), Hashmi (1973), and later by Subramanian (1971), and others have recognised it as basauxic conidiophores and interpreted that its development is like in *Spegazzinia, Arthrinium* and others. *Dictyoarthrinium* is a tropical fungus and it has been collected from in late winter and summer seasons. *Dictyoarthrinium* grows in culture and it shows intercalary growth.

15. DICTYOCHAETA Spegazzini

Physis, 7: 18, 1923

Dictyochaeta was proposed by Spegazzini (1923), with *Dictyochaeta fuegiania* Spegazzini, as its type species. It produces effuse colonies with or without setae. Mycelium partly superficial, partly immersed, hyphae septate, branched, forming compact bodies similar to stroma. Conidiophores macronematous, mononematous, solitary or in groups like fascicles. Conidiogenous cells polyphialidic, terminal or intercalary. Conidia phialidic, hyaline, elongated with or without setulae, one or on either side.

Dictyochaeta is cosmopolitan foliicolous fungus. It has been reported from New Zealand by Hughes and Kendrick (1968), Kuthubutheen and Nawawi (1991), have reported number of species from Malaysia. Kuthubutheen and Nawawi (1991), have proposed key to 59 species of this genus. However, Bilgrami et al. (1991), have not taken this genus into account while indexing the Indian fungi. Holubova – Jechova (1988), has reported many species from Czekoslovakia and other European countries. Mercado et al. (1997), have recognised 3 species from Cuba and illustrated them beautifully. Bhat and Kendrick (1993), described *D. ciliata* and *D. tropicales* from Goa, Karnataka and Andaman Islands, India.

Dictyochaeta parva (Hughes & Kendrick) Holubova-Jechova, Ceska. Mycol Praha. 42: 204, 1988. (Fig. 17)

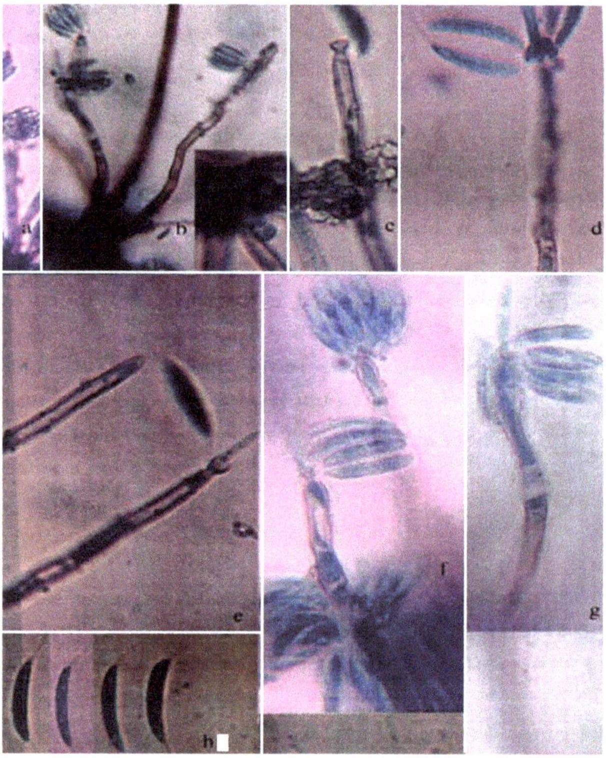

Fig. 17. *Dictyochaeta parva* - a-b, Colony (450 x).
c-g, Conidiophores, conidiogenous cells with conidia (1000 x)

Colonies effuse, brown with whitish pin heads. Mycelium superficial or incorporated into weak stromatic body with thick walls, 3-7μm broad, sometimes chalmydospore like bodies present. Conidiophores macronematous, mononematous, arising in tufts or solitary, laterally from the creeping hyphae, 80-100μm long, septate, cylindrical. Conidiogenous cells phialidic, terminal or lateral or intercalary with a distrinct collarette, collar, 2-4μm wide.Conidia phialidic, crescent shaped, hyaline, thin walled, aggregated into mucilaginous mass for sometime and later on getting separated, 10- 15μm long, 2-3μm broad, pointed at both ends and provided with fine erect, setuale, 2-7μm long. No secession and germination were observed.

On leaf litter, debris of plant material from Adilabad, Karimnagar, Srikakulam districts of A.P., HCIO No: 43,008. NCBL No. 161, 213. Sep. 2001. Coll. V.R.&G.S.K.

Dictyochaeta in the latest sense is being recorded for the first time in India and earlier it was reported by other workers as *Codinaea*. Holubova-Jechova(1988), has transfered almost all the species of *Codinaea* into *Dictyochaeta* because it antedates publications. *Dictyochaeta* was published in 1923 and later *Codinaea* was proposed by Maire (1937), which has similar characters. Therefore, *Dictyochaeta* is valid and accepted name by all mycologists. This fungus is a foliicolous fungus usually found and reported from western ghats of India and Assam.

16. DIDERMA Persoon

Neu.Mag.Bot.1:89, 1794.

The genus *Diderma* was created by Persoon in 1794, with *D. globosum* as its type species. He included 4 species namely *D. floriformae, D. globosium, D. contortum* and *D. deformae*, but later workers (Lakhanpal & Mukerji, 1981), found that they are doubtful species. *D. globosum* is the type species of the genus. It produces fruit bodies which are sporangiate, sessile or stalked sometimes plasmodiocarpus are even pseudothalliod peridium clearly bears double walled, calcareous or cartilaginous often shining capitulum of branched, anastomosed, limeless threads or sometime replaced by slimy columns. Spores rounded, globose, dark coloured in mass.

Lakhanpal and Mukerji (1981), have given an account of this genus and proposed a key to the known species in India. They have described about 24 species.

Diderma globosum Pers. Neuse. Mag. Bot. 1: 89, 1794. (Fig. 18)

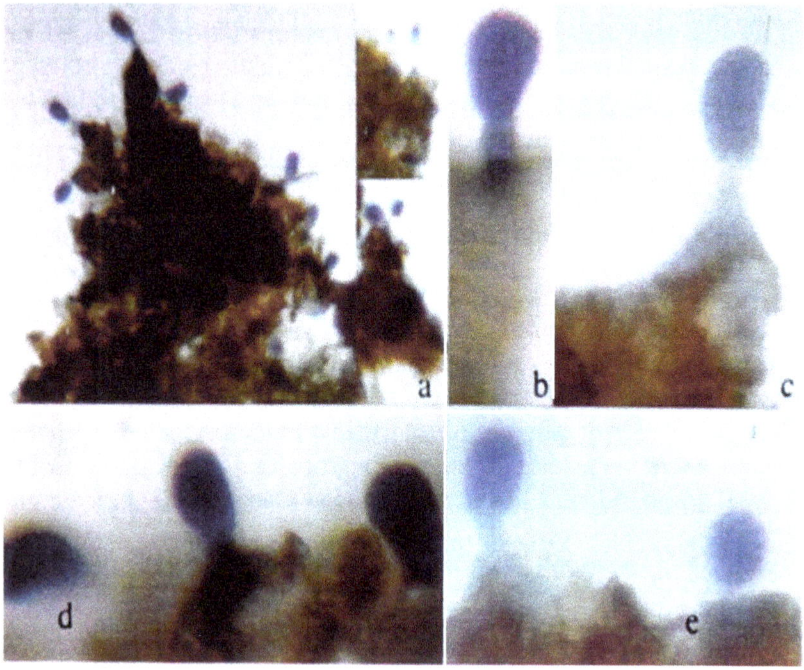

Fig. 18. *Diderma globasum* - a, Colonies (100 x). b-e, Fruit bodies (450 x).

Fructification sporangiate, sometimes subplasmodiocarpus. Sporangia slightly stalked to sessile, 1-2μm in diam. Assuming pseudothallial proportions whitish in colour, hypothallus usually profused, peridium double walled, outer layer smooth, calcareous, inner layer smooth wrinkled in the specimen of Khammam. Collumella and spores did not develop, the fruit body was still in pseudo or subplasmodial stage.

On decaying wood logs from Medak, Karimnagar, Bhadrachalam and Yellandu forests of Khammam district, A.P., HCIO No: 43,009. NCBL No. 12, 76. Sep./Oct. 2001. Coll. V.R. & G.S.K.

D.globosum was collected in N. India by Lakhanpal and Mukerji (1981), and Agnihothrudu (1961), from Assam, India.

17. GANODERMA P. Karsten

Symbolae Mycolo. Fennicana 33 parts, 1881

Ganoderma belongs to Basidiomycotina, Ganodermataceae (aphyllophorales) and has over 50 species described in the tropical countries. *G. applanatum* and *G. lucidum*, cause wood decay and *G. philippii* is responsible for root rot. They are common and well known species occur in all parts of the world. The genus is identified by its basidiocarp which is 5-50 *cm* across, sessile, horizontal,semicircular, bracket shaped, flatish but with humps that are more are less concentric, rising to the point of attachment, smooth, dull reddish brown or coccova coloured from the deposited spores.Spores small whitish turning brown than cinnamon and finally brown with age. *Ganoderma* occurs on trunks of fallen or aging trees, it causes rotting of wood on timber and forest wood stacks. Peterson(1987),has reviewed this genus with 5 species and provided a key to its species and appended its relevant literature. According to Bilgrami et al. (1991), 5 species of *Ganoderma* have been reported from India.

Ganoderma applanatum Fries (Karsten) in Soothill and Fairhurst, New Field Guide to Fungi. P. 68, 1978. (Fig. 19)

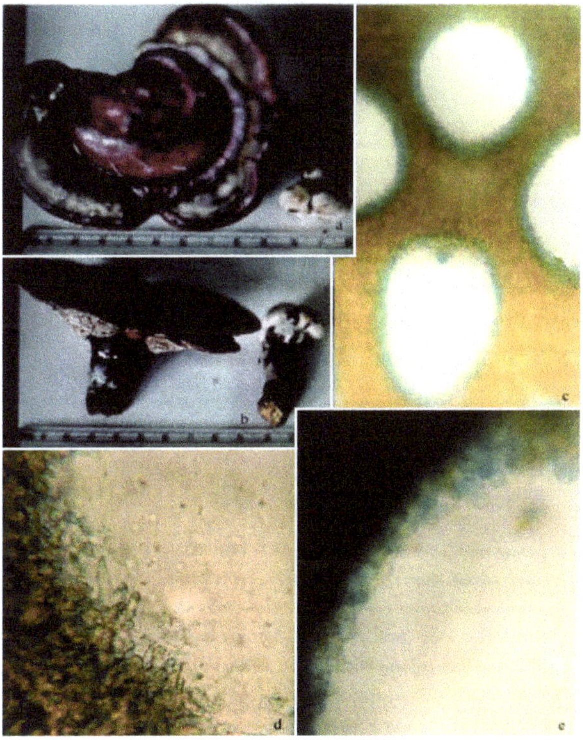

Fig. 19. *Ganoderma applanatum* - a-b, Basidiocarp and its button stage, c-e, V.S. of basidiocarp basidia and basidiospores (1000 x)

Ganoderma basidiocarps 5-10 *cm* in diam., reddish coloured, disc-like semicirculer, bracket shaped, flattish strange hard, smooth or dull reddish brown or coccova coloured from the deposited spores. Mycelium incorporated in the woody substrate thin walled hyphae or thick walled sclerotic cells may be present. Sectional view shows pores the basidia and cystidia that occur in the open cavity. Basidia are red in colour, ornamented when young, whitish later cinnamon and brown with age.

On decaying wood logs from Bhadrachalam and Ashwaraopet forests of Khammam district, A.P., HCIO No:43,048. NCBL No. 28, 118 Nov./Dec. 2001. Coll. V.R. & G.S.K.

Ganoderma is a wood rotting fungus but according to some scientists it has been employed as filter for effluent treatment in which toxic substances and high alkaline or acidic chemicals are observed which are left untreated. The filters made by *Ganoderma* and *Fomes* have been good and the effluent water treatment resulted with above fungi in bio-remediation.

18. GONYTRICHUM C.G. & T.F.L. Nees ex Wallroth

Flora Krypt. German. 2: 274, 1833

Gonytrichum was proposed by Nees and Nees in 1818 (Nova. Acta. Acad. Caes. Leop. 9: 244), with *G. caesium* as its type species. *Gonytrichum* produces effuse colonies with septate mycelium. Conidiophores macronematous, mononematous and branched. The branches ultimately becoming setiform and coiled. Conidiogenous cells phialidic bearing in clusters on the axis of the conidiophores or at the junction of the branching. Conidia phialidic arising from multiple conidiogenous loci form whitish head, cylindrical, hyaline and guttulate one celled.

Subramanian (1971), described *G. microcladium* from India, other species of *Gonytrichum* are reported to have come from U.S.A. Canada, Europe, England and India. Gams and Holubova-Jechova (1976), have studied many species of *Chloridium* and *Gonytrichum* growing on decaying food. They have described number of varieties and new species. *G.microcladium* (Saccardo) Hughes, was described from India by Subramanian 1971, which produces setae and branches in the apical region and conidiogenous cells were confined to lower little part.

Gonytrichum microcladium (Sacc.) Hughes, Trans. Brit. mycol. Soc. 34: 565, 1951. (Fig. 20)

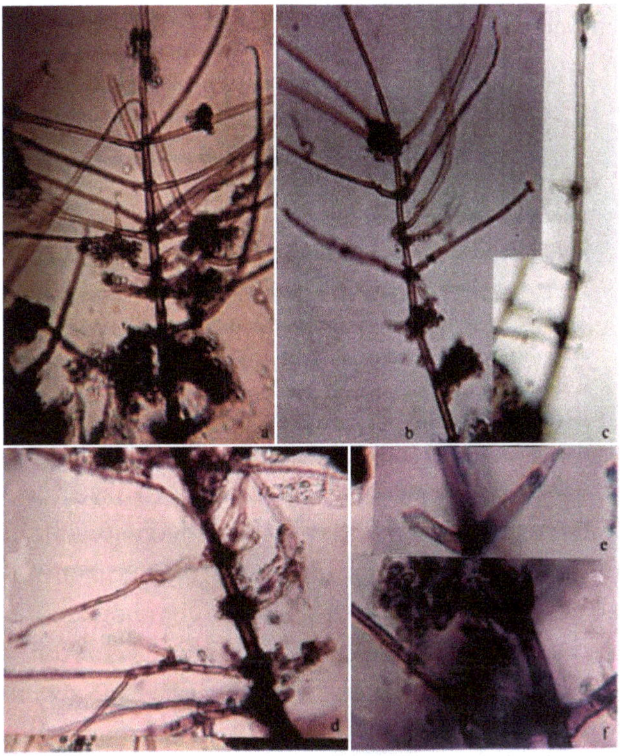

Fig. 20. *Gonytrichum microcladium* - a-f, Conidiophores with setae, conidiogenous cells and Conidia. (a-d 450 x, e-f 1000 x)

Colonies effuse olivaceous brown, branched hairy. Mycelium immersed, septate, 2-3μm wide, sub-hyaline to brown, smooth, branched. Conidiophores crowded on the substratum but arising in single, macronematous, mononematous, erect, 250-350μm long, 5-8μm wide at base, 4-5μm wide and tapering at apex, 1-1.5μm wide, some part of the based region, thick walled, smooth after lower quarter of the conidiophore. Conidiogenous cells are produced in the apical part consisting of branches which may be tapering subulately and becoming paler towards the end. These branches may be sterile or fertile confined to apical region or sometimes mixed with the conidiogenous cells itself. Conidiogenous cells arising on the main axis of the conidiophore(Stipe). Simple or in whorls of 3-4 divergent. The phialides are variable in length from 10-20μm long, 3-4 wide, sub-hyaline to pale brown collariate at the tip divergent to the main axis. The setae when present tapering subulately to form a hyaline often circinate or curved tip,

70-170μm long directed upwards smooth or varrucose. Conidia oval, one celled, hyaline, enteroblastic, sub-hyaline to very pale olivaceous brown, 3-4μm long, 2-2.5μm broad, in slimy mass at the apex of the phialide or in straight or curved columns of aggregated conidial mass or the conidia occur in groups of two and three.

Collected from debris and leaf litter from Srikakulam, Vishakapatnam, Bhadrachalam, Yellandu forests of Khammam district, A.P., HCIO No: 43,010. NCBL No. 23, 137 Nov. 2001. Coll. V.R. & G.S.K.

Gonytrichum isolated collected in Khammam forest is slightly strange when compared with *G. caesium*. The conidiophore is zig-zag and the setulose branches are confined to mixed regions. In *G. microcladium* the setae are confined to the apical part of the conidiophore. In this collection it has been observed that the conidiogenous cells are of variable in shape and are distributed any where indiscriminately. It has been observed that there is a tendency of the setae to turn into fertile conidiogenous cells, therefore, the conidiogenous cells are very long. It may suggest modification of conidiogenous cells into setae or transformation of satae into conidiogenous cells. It could not be compared with other collections of India. Therefore, the name *G. microcladium* is given which was originally described by Subramanian (1971), who isolated *G. microcladium* from soil and *Dalbergia* forest in Uttar Pradesh.

19. GYROTHRIX Corda

Icones Fungorum, 5: 13, 1842

Gyrothrix was proposed by Corda (1842), with *G. podosperma* (Corda) Rabenhorst, as its type species. *Gyrothrix* is a foliicolous fungus. It is characterised by punctiform to effuse, grey to black or dark brown, velvety colonies. Mycelium superficial or immersed in the substratum, septate, branched, smooth walled bearing setae and conidiophores. Setae arising from the mycelium which is repeatedly branched, erect, straight or flexuous, septate, thin walled, smooth or verrucose, broad at the base and tapering at the apices. Conidiophores macronematous or semi-macronematous similar to conidiogenous cells which are obclavate to lageniform, thin walled. Conidia phialidic (enteroblastic) or aggregated into a whitish layer at the base of the setae, acerose to fusiform, straight or curved, one celled smooth.

Gyrothrix is a mitosporic fungus and there are about 15 species widely distributed in tropical countries as a saprophyte on leaves of different plants. Pirozynski (1962), presented an illustrative account of 12 species from CMI Kew England. His studies were based upon the natural substrates and there was a doubt about its conidiogenesis. Cunningham (1974), revised the genus and gave an illustrated account and a key to its species and proposed a new species based upon the cultural characters. Pirozynski and Patil (1970), described some species from India and discussed their ecological aspects. *Gyrothrix* was known to India by its 11 species described by various workers from different states of India. Vasant Rao and de Hoog (1986), proposed *Gyrothrix cornuta* from Karnataka. The genus was also described by Matsushima (1971), on several occasions from Japan, Solomon Islands and other places.

Gyrothrix podosperma (Corda) Rabenhorst,Deuschl.Krypt.Fl. 1:72,1844. (Fig. 21)

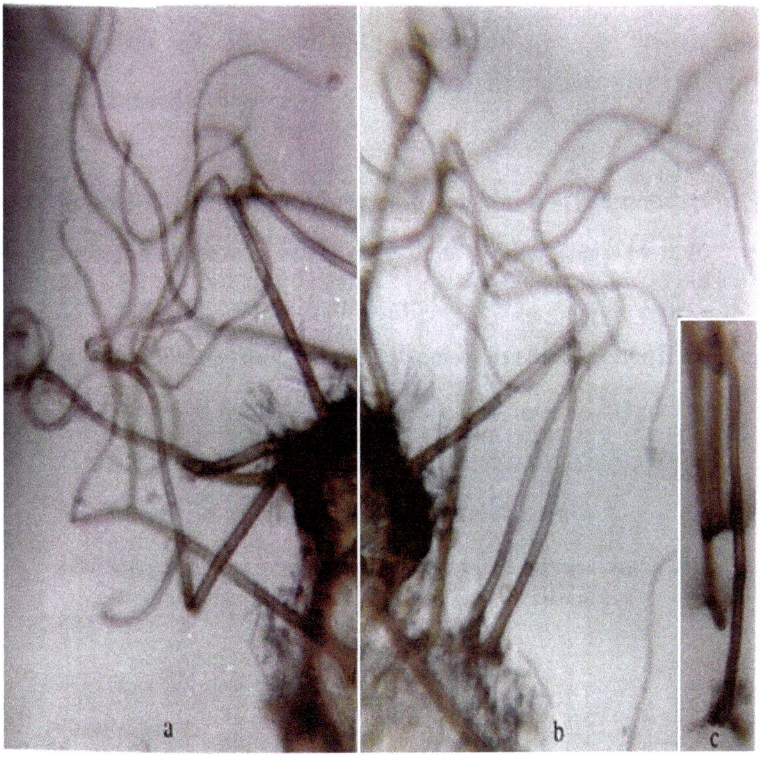

Fig. 21. *Gyrothrix podosperma* - a-c, Colonies with setae, foot cells, conidiophores, conidiogenous cells and conidia (450 x).

Colonies effuse, dark reddish brown, velvety, irregular, 1-5 *mm* in diam., scattered or united. Mycelium forming a net work on the epidermal walls and cuticle of the host, superficial hyphae branched, anastomosing, smooth, sub-hyaline to brown, bearing setae and conidiophores. Setae numerous, intertwined, dark brown at the base, rough, thick walled, 4-6 times sub-dichotomously branched, 100-250μm in length. Trunk of the setae straight or flexuous, 50-150μm in length, 2.5-4μm wide at the base above the bulbous swelling, smooth or minuetly rough, branches primary, secondary and tertiary types tapering at the apex with a width of 1 *mm* broad. The tips may be circinate, spirally twisted, rough, colourless thin walled. Conidiophores similar to conidiogenous cells, borne laterally on superficial hyphae obclavate to lageniform, sub-hyaline, 5-10μm in length, 3-4μm broad at the base and upto 1.0μm broad at the apex. Conidia enteroblastic, phialidic aggregated into whitish layer at the bases of the seta, cylindrical to fusicoid, straight of curved with the attached end to the phialide subulate and the free and obtuse continuous, smooth or one celled, hyaline, 7-14μm long and 1-2μm broad.

On dead leaves from Bhadrachalam, Yellandu forests of Khammam, Nizamabad, Srikakulam districts of A.P., HCIO No: 43,011. NCBL No. 145, 225. Oct. 2002. Coll. V.R. & G.S.K.

Gyrothrix reminds *Circinotrichum* and like genera which bear simple or branched setae, conidiophores or conidiogenous cells on the creeping hyphae or under the setae itself. Kendrick (1980), suggested to merge these genera into one tribe as in the case of *Beltrania* and others. The criteria for this suggestions are the setae and conidiogenous cells, conidia, the foot cell and the setae. There is no much cultural work on this genus, but it seems that it degrades the leafy matter and helps the formation of humus in the soil and adds to its fertility to support or to sustain on other angiospermic plants.

20. HANSFORDIA Hughes

Mycol. Pap. 43: 15-24, 1951

Hughes (1951), proposed *Hansfordia* with *H. ovalispora* as its type species and other six more species namely *H. togoensis, H. grewiae, H. ugandensis, H. triumfettae, H. canescens* and *H. gricella*. Colonies effuse, small, tree like, whitish grey and hairy. Mycelium immersed or superficial.

Conidiophores macronematous, mononematous, erect, branched, like primary, secondary, tertiary branches. Conidiogenous cells terminal or intercalary on the terminal branches, polyblastic, cicatrised, geniculate. Conidia holoblastic, acropleurogenous, denticulate with truncate base with conical apex, smooth, spherical or oval.

While proposing the genus, Hughes (1951), recorded that it occurs as a mycoparasite also. It is also found as a saprophyte on dead leaves of sugarcane, in Togoland, Gold Coast colony of S. Africa, Uganda, Argentina and other places including India. *H.cupilifora* was observed as a parasite of *Ginnosporia* leaves in Karnataka and *Cassia* leaves in Colaba, Maharashtra state.

Hansfordia griciella (Sacc.,) Hughes, Mycol. Pap. 43: 16-24, 1951.
(Fig. 22)

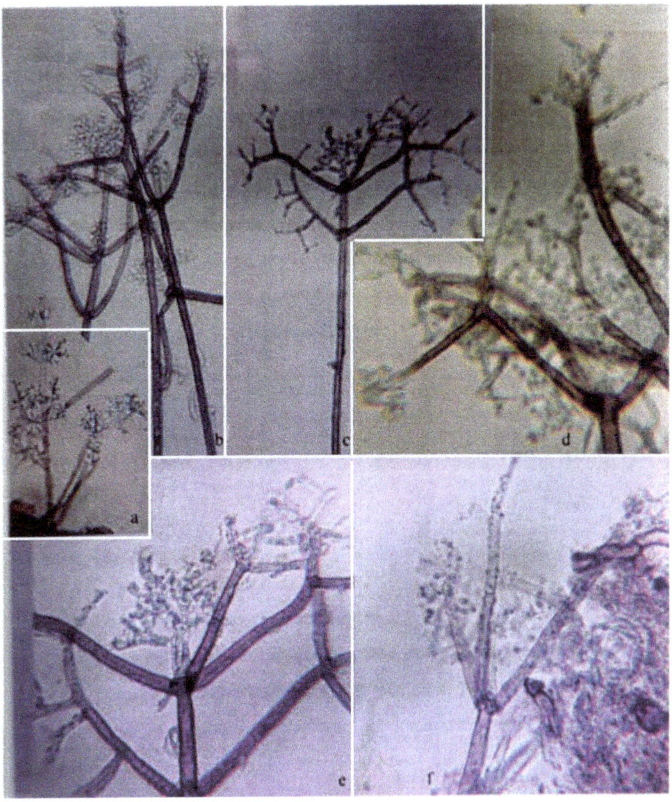

Fig. 22. *Hansfordia griciella* - a, Colony (100 x)
b-f, Conidiophores, conidiogenous cells and conidia (b 450 x, d-f 1000 x)

Colonies effuse, pin head like, whitish grey, dispersed. Mycelium superficial, thick walled, septate, brown, 3-4*μm* broad, smooth. Conidiophores macronematous, mononematous, arising from repent hyphae, erect, straight, branches in the terminal region, septate, thick walled, pale to blackish brown, 3-7*μm* broad at base, 40-150*μm* long, 2-4*μm* broad at the apical region. Conidiophores branched in the apical region, characteristically, intercalary cells or whorls of branches or of the order like primary, secondary or tertiary. Conidiogenous cells terminal or intercalary, polyblastic, cicatrised, geniculate, 15-30*μm* long, 2-3*μm* broad at base, 1-2*μm* broad at the apex, scars distinct and flat. Conidia polyblastic, acropleurogenous, one celled, oval, 1-3*μm* diam. Colourless or whitish. No conidial germination was observed.

On decaying plant debris and leaf litter from Bhadrachalam, Yellandu forests of Khammam, Warangal, Nizamabad districts of A.P., HCIO No: 43,012. NCBL No. 155, 332. Oct./Nov. 2002. Coll. V.R. & G.S.K.

21. HELIOCEPHALA Vasant Rao et al.

Persoonia 12(3): 239-242, 1984

Heliocephiala was created by Vasant Rao et al. in 1984 with only one species i.e *H. proliferans* Rao et al. as type species from India. *Heliocephala* is well characterised by radially arranged, pale brown, borne from monoblastic ampulliform conidigenous cells that are characterised at the top of slender, micronematous conidiophore. The conidia germinate by their apical rostrate cell to develop secondary cluster of conidiogenous cells which bear conidia again. The conidia are three celled with rough basal cell and smooth upper cells which ultemately become curved or uncinate.

Heliocephala remained monotypic till the discovery of another species namely *Heliocephala zimbabwensis* (Decock et al. 1998). The second species was collected on unidentified rotten leaf at fuller forest in Zimbabawe it was well studied in culture and in scanning electron microscope. Recently another new species was collected from Tamilnadu and named *Heliocephala natarajanai*. In the present study a specimen was collected from Tirumalagiri, dt. Nalgonda, which happens to be similar to *H. proliferans* Rao et al. is being described.

Heliocephala proliferans Vasant Rao, Reddy & de Hoog, Persoonia 12(3):239, 1984. (Fig. 23)

Fig. 23. *Heliocephala proliferans* - a, Colony (100 x). b-c, Conidiophores showing proliferation (450 x).

Colonies effused light grey formed of erect individual conidiophores with radiating conidial heads which may produce secondary and tertiary conidial heads. Mycelium superficial and immersed hyphae brown, thick walled, 3-5μm broad, septate. Conidiophores macronematous, mononematous, determinate erect, straight, unbranched, brown, 3-7 septate, thick walled, 150-300μm long, 3-4μm wide, conidiogenous cells in clusters holoblastic, monoblastic, discrete, ampilliform, thin walled, smooth, hyline to brownish, 5-6 x 3-7μm borne successively new conidiogenous cells arisisng laterally on old ones and then forming compact cluster at the top of the conidiophore. Conidia blastic, arising radially inheads ellipsoidal, rostrate, slightly obovoide, 25-100μm long, 3-5μm broad, rostrum short to very long, straight or curved at top cap 75μm long and 2μm broad, and taparing at apex. Conidia 2 celled, septate, slightly smooth conidia germinate through there rostrum which produce secondary cluster of conidiogenous cells.

Collected on unidentified rotten leaf in water bodies at Tirumalgiri, Nalgonda, Karimnagar, Khammam, A.P., NCBL No. 285, 324. Nov. 2002. Coll. V.R. & G.S.K.

It seems Dr Decock and V.Robert from University of Louvain, Belgium, tried to contact Dr. Vasant Rao and to examine the type material which was deposited at CBS Baarn, Holland, and Vivek Vardhani college,.Hyderabad, A.P., India, and he left both the places. The type of material was not available. It is unfortunate that the matarial could not be sent to Dr. Decock & V.Robert for reasons beyond control, in the present study also we could not examine the type material and material subsquently published for the later species *Heliocephala* is a follicolus anamorphic fungus and its significance could not be established nor there could be any biochemical investigations despite the fact it grows well in culture.

24. KINNERASANIOMYCES G.S.K. & V.R.Gen. et sp. nov.

Kinnerasaniomyces Gen. nov.

Deuteromycotina, Agonomycetales (Mycelia Sterilia), Agonomy- cetaceae. Colonies effuse, stout, visible to naked eyes, dendroid, branched at the apex. Mycelium immersed, septate, branched. Conidiophores, macronematous, synnematous, erect, flexuous, branched. Conidiogenous cells synchronous, sympodial, terminal or lateral. Conidia bullet shaped, synchronous, multicellular formed by group of hyphae, rounded at apex, cylindrical, abruptly flat at base, consisting small stalk made up of several hyphae. Conidial secession rhexolytic.

Type species - ***Kinnerasaniomyces elagans***

(Etym - Kinnerasani = place of collection and myces = fungus, elagans = beautiful).

Kinnerasaniomyces Gen. nov.

Deuteromycotina, Agonomycetales, Agonomycetaceae. Coloniae effusae,dendroidae, ramosae ad apicem. Mycelium immersum, septati, ramis. Conidiophora macronemata, synnemata, recta, flexosa, rami. Cellae conidiogenae synchronocio, sympodicae, terminalae vel laterribus. Conidia formentibus anamolae, synchronocio, modous bulleti pellatae formae, multicellulata, cylindricae, rotendibus apicem, regiobasialae flattendibus adpedicellata,secessio rhexolyticae.

Species typica - ***Kinnerasaniomyces elagans.***

Presently it is monotypic and being reported from India for the first time.

Kinnerasaniomyces elagans G.S.K. & V.R. sp. nov. (Fig. 24)

Fig. 24. *Kinnerasaniomyces elagans*
a-d, Colonies with conidiophores and propagules (a-b 100 x, d-e 450 x)

Coloniae effusae, brunnei auribrunnei, rami. Mycelium immersum, ramis septatis, sub-hyalina vel brunnei, 2-4μm lata. Conidiophora recta, macronemata,synnemata, curvata, textura angularis presumatica vel porecta. 300-1000μm longa, 30-70μm lata ad basim usque 20μm lata ad apicem. Cellae conidiogenae terminale, integrata, synchronicio, formentibus, anamolae. Conidia formentibus hyphis fasciculata, bullet formae, apicem rotenda, cylindricae, basim flatendibus vel pedicellata formentibus, fasciculata, hyphis 50-90μm longa, 20-30μm lata, auribrunnei. Conidial secessio rhexolyticae.

Bhadrachalam forestae, Khammam district, A.P., HCIO No : 43,031. NCBL No. 368. Nov. 2002. Coll. V.R. & G.S.K.

Kinnerasaniomyces elagans G.S.K. & V.R. sp. nov.

Colonies effuse, yellowish brown, branched, stout. Mycelium immersed, hyphae branched, septate, sub-hyaline to brown, 2-3μm broad, immersed in the host tissue. Conidiophores macronematous, synnematous, straight or curved, surface of the synnemata smooth, textura angularis presumatica or porecta. 300-1000μm long, 30-70μm broad at the base, up to 20μm broad at the apices. Conidiogenous cells terminal, integrated, synchronous, conidia formed by group of hyphae. Conidia bullet shaped, synchronous, conidiogenesis unknown, but on one occassion it was observed that a globular mass disintegrates into small pieces still attached on the tip of the branch into numerous segments which get transformed into conidia like propagules. The conidia rounded at the apex, cylindrical, flat at the base having a multi-hyphal fractured stalk. 20-30μm broad, 50-90μm long, pale yellowish brown. Conidial secession rhexolytic. No germination was observed.

From Bhadrachalam forest of Khammam district, A.P., HCIO No:43,031. NCBL No. 368. Nov. 2002. Coll. V.R. & G.S.K.

Agonomycetales is a group of fungi which contain about 18 genera. They do not produce true conidia, but dehiscent propagules (allocysts, bramatia, bulbils, chlamydospores or sclerotia) like bodies are produced in this group of fungi. Representing Basidiomycotina or Ascomycotina or other mitosporic fungi like *Rhizoctonia, Sclerotia, Papulaspora, Aegerita, Bulbilomyces, Tretopileous, Mycoenterolobium, Arbuscula, Cancelidium, Arachilispora, Brugoa, Minimedusa* and some of them are plant pathogenic where as others are ectomycorrhizal or so. Their biology and taxonomy remains unsettled and incomplete. Taxonomically these genera are lumped into mycelia sterilia, because there is no definite mode of conidiogenesis. Carmichael et al.(1980),gave an account of about 13 genera, whereas Hawksworth et al. (1995), stated more than 28 genera with their synonyms covered 30 and 200 species. It seems that there is no genus to accomodate the above fungus. The synchronous mode of conidiogenesis is comparable to *Arbuscula, Tretopileous* and others. In *Arbuscula* the bulbils are formed in acrogenous chains, whereas in *Tretopileous* the conidiomata produces disk like propagules which get separated in others bulbil forming members which may be hollow or solid or sometimes may contain cavities producing pycnoconidia. The name Kinnerasaniomyces is given after the place of its collection. Kinnerasani project in Khammam district which has a natural vegetation, the specific epithet elagans means beautiful.

23. LATERICONIS Vasant Rao et al.

Mycotaxon, 19: 409-412, 1984

Latericonis was created by Vasant Rao et al. with *L. obscura* as its type species . *Latericonis* produces effuse, dark to blackish brown colonies, Septate mycelium, conidiophores macronematous, mononematous, geniculate, conidiogenous cells monoblastic, termianal, integrated, intercalary, becoming sympodic bearing holoblastic, fusiform, tangentially produced muriform conidia. It was originally collected from Orissa, India. In the present investigation, on unidentified bark, a similar taxon was isolated which is being described here.

Latericonis obscura Vasant Rao et al. Mycotaxon 19 : 409 – 412, 1984. (Fig. 25)

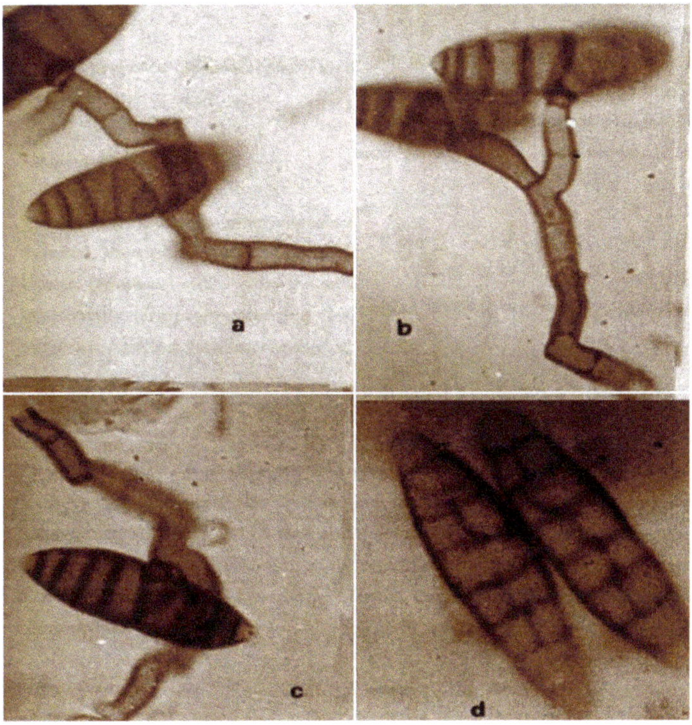

Fig. 25. *Latericonis obscura*
a-d, Conidiophores with conidiaferous cells and conidia (1000 x)

Colonies effuse, dark to blackish brown. Mycelium immersed, visible, hyphae septate, 2-5μm broad, smooth, pale brown. Conidiophores macronematous, mononematous, simple, flexuous, ziz-zag or geniculate, septate, cylindrical, smooth, pale to dark brown, 30-70 x 3-4μm, tapering at apex. Conidiogenous cells terminal, monoblastic, intercalary, sympodic growth very characteristic i.e. the next conidium arise due to helical growth of the conidiophore. Conidium holoblastic, terminal, transverse, obliquely and vertically septate, fusiform, rounded or pointed at wither either ends, 25-50x 10-15μm, rhexolytic conidial secession.

Collected on unidentified bark, Dindi river, Nalgonda, Srikakulam, Visakhapatnam, A.P. NCBL No. 33, 149. Nov. 2001. Coll. V.R. & G.S.K.

Latericonis was originally reported from Orissa. It is the second collection from Dindi river project, Nalgonda Dt. It is quite possible that this taxon can be located in Eastern and Western ghats of Indian forests. It is strange that the septation in Nalgonda collection is transverse, oblique and verical. This fact was not observed earlier.

24. MELANOCEPHALA Hughes

Newzealand J.Bot. 17: 166-173, 1979

Melanocephala was introduced by Hughes (1979), with *M. cupulifera* as its type species. In the same communication he added *M. australiensis* (Shearer, Crane & Miller) Hughes. *Melanocephala* produces black effuse colonies. Mycelium immersed. Conidiophores macronematous, mononematous, simple, straight. Conidiogenous cells sub-cylindrical or inflated, proliferating and percurrent. Conidia holoblastic, broadly ellipsoidal to ovoid 2-3 septate, secceeding rhexolytically from the conidigenous cells which is the distinguishing point from *Endophragmia.* Duvernoy et Meyer and *Endophragmiella* Sutton. Hughes (1979), has discussed the number of allied genera and he pleaded for free location of species of *Endophragmia* giving notes under relavent generic names.

Melanocephala with its three species was known from New Zealand, Canada, Australia and U.S.A. Later, Kirk (1983), Vasant Rao and de Hoog(1986), reported the same species from Africa and India respectively.

Melanocephala cupulifera Hughes, New Zealand J. Bot. 17: 168-169,1979. (Fig. 26)

Fig. 26. *Melanocephala cupulifera* - a-l, Colony, Conidiophores with percurrent conidiogenous cells and conidia (a 100 x, b-i 450 x, j-l 1000 x)

Colonies effuse, dark. Mycelium immersed, visible hyphae septate, branched, sometimes forming a small aggregation of hyphae with thick walled variable cells resembling to a weak stroma. Stroma made up of loose thick walled cells penetrating the substrate or anchoring the conidiophore with rhizoid like hyphal branches. Conidiophores macronematous, mononematous, straight, erect or curved, cylindrical, dark, thick walled, 80-100μm long, 5-7μm broad at base, 5-9 μm broad at apex. Conidiogenous cells terminal, cylindrical at base and than expanding into a conical or cup shaped distal part 10-15μm long, 7-12μm broad, proliferating percurrant upto 4 percurrations, monoblastic, integrated.Conidia monoblastic, broadly ellipsoidal 3-4 septate with lower cell usually shortest and less pigimented as the conidia become mature the septa become markedly deep pigmented on surface, 30-45μm long, 15-20μm broad and 7-10μm broad at the base, truncate, the conidial secession rhexolytic. No germination was observed.

On decaying wood from water bodies in Warangal, Karimnagar, Bhadrachalam, Yellandu forests of Khammam district, A.P., HCIO No:43,016. NCBL No. 61, 173. Nov. 2001. Coll. V.R. & G.S.K.

25. MEMNONIELLA Hoehnel

Zen. Bakt. Parasi. Abt. 2,6: 16-17, 1923

Memnoniella was proposed by Hoehnel in 1923, with *Memnoniella aterrima* Hoehnel, which was later observed to be similer to *M. echinata* (Riv.) Galloway. *Memnoniella* produces mycelium rarely septate, branched. Conidiophores macronematous, mononematous, simple straight, septate. Conidiogenous cells phialidic with one or two whorls at the tip of the conidiophores. Conidia phialidic, one celled in long chains.

Mamnoniella and *Stachybotrys* were considered to be synonyms by Carmichael et al. (1980). However, Simth (1962), Jong and Davis (1976), and Ellis (1976), have still treated them separately. The point of difference between the two is only the presence and absence of conidial chains. Subramanian (1971), reported two species from India with *M. levispora* Subram., as new species. Hawksworth et al. (1995), recognised 4 species of this genus. From Cuba, Mercado et al. (I 997), described all of them with a new synnematous *M. laprosa* of Castaneda.

Memnoniella echinata (Riv.) Galloway, Trans. Brit. mycol. Soc. 18:163-166,1933. (Fig.27)

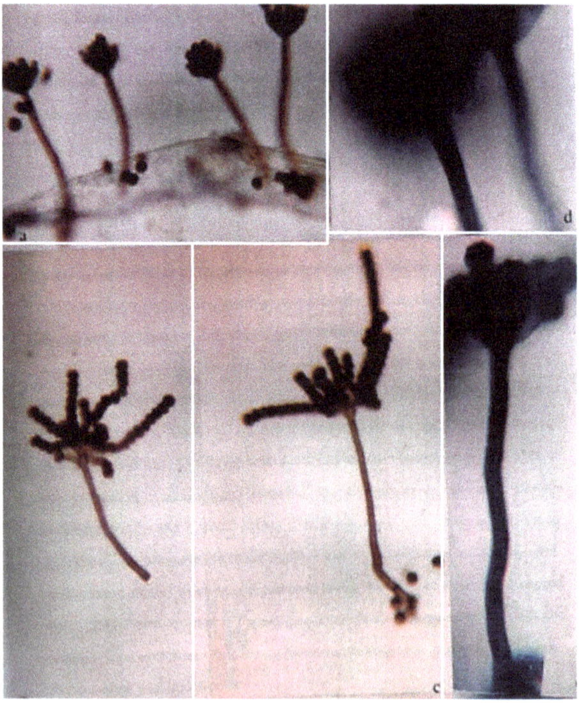

Fig. 27. *Memnoniella echinata*

a-e, Colony, conidiophores with phialides and conidia (450 x, 1000 x)

Colonies disperse, black, pulvinate, powdery occupying several millimeters on the substratum. Mycelium superficial or immersed in the substratum, septate, branched, hyaline to sub-hyaline, smooth. Conidiophores, macronematous, mononematous, septate, cylindrical slightly darker than the mycelium, smooth or verrucose, 60-100*μm* long, 3-4 *μm* broad. Conidiogenous cells phialidic, phialides 6-9 *μm* long and 3-5 *μm* broad. Conidia arising from the phialides spherical, verrucose in long chains but getting separated slowly into ones or twos, 3-5 *μm* in diam. No germination and conidial detachment was seen.

On decaying and wet leaves, twigs and barks etc., from water bodies in Warangal, Bhadrachalam, Ashwaraopet, Yellandu forests of Khammam district, A.P., HCIO No:43,017. NCBL No. 44, 193, 241. Nov. 2001. Coll. V.R. & G.S.K.

Memnoniella echinata is collected in post rainy seasons as it prefers more moisture and cellulosic substrates.

26. MENISPOROPSIS Hughes

Mycol. Pap. 48: 59-61, 1952

Hughes (1952), created *Menisporopsis* with *M. theobromae* as its type species collected on leaf litter of *Theobroma* in New Zealand. It is recoginised by colonies which are hypophyllous, dispersed, greyish pink with stout projecting setae masked by the conidial head. They may sometimes over grow profusely through the conidial mass covering the setae and over growing into a rounded catkin like appearance. Mycelium superficial, septate, brown coloured, branched, thick walled stroma. Stroma slightly raised above the sub-stratum, pseudoparenchymatous. Conidiophores macronematous, synnematous. Conidiophores branched many times, erect, flexuous, closely adpressed with the central setae. Conidiogenous cells monophialidic or polyphialidic, integrated terminal cylindrical with collarette. Conidia semiendogenous, phialidic, cylindrical, fusiform, curved, hyaline, rarely 0-1 septate and usually with or without appendiculated, setuliform on extreme ends, the conidia aggregated into mucilaginous mass covering setae partly or completely. Setae central septate, erect.

Menisporopsis was again reported by Hughes and Kendrick (1968), from New Zealand after its creation. Matsushima, reported *M. theobromae* and *M. neiwzealandae* from Japan and Solomon Islands. Pirozynski and Hodges

(1973), created *M. profusa* a Canadian species. Kirk and Sutton (1985), observed *M.ludoviciana* (Crane and Schoknecht) Kirk and Sutton, as an another species. Varghese and Rao (1978), proposed *M. pirozynski* as a new species from India. Recently, Vasant Rao and de Hoog (1986), proposed *M. pleiosetosa* from India, which was also collected by Matsushima (1996), from Malaysia. Matsushima (1996), preferred to recognise 5 species of this genus. He illustrated *M. pleiosetosa* with phase contrast microscopic photographs.

Menisporopsis theobromae Hughes, Mycol. Pap. 48: 59-61, 1952.

(Fig. 28)

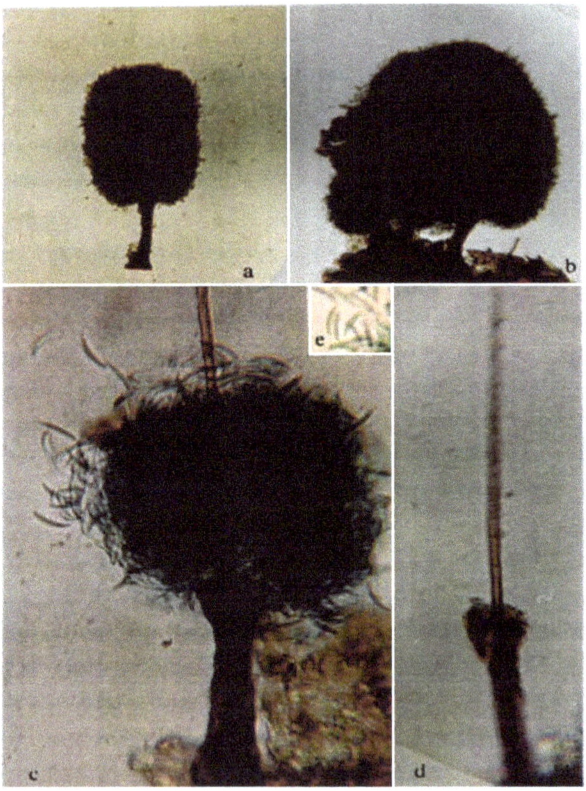

Fig. 28. *Menisporopsis theobromae* - a-e, Conidiomata, conidiophores setae, conidiogenous cells and conidia (a 100 x, b-e 450 x)

Colonies effuse, cotton plug like, whitish. Mycelium partly superficial, partly immersed, thick walled, pale to dark brown, 3-5μm broad, septate, stroma pseudoparenchymatous, erumpent, thick walled, producing conidiophores and setae. Setae stout, septate, cylindrical, erect, unbranched, curved, thick walled upto 350μm long, 5-7μm broad at base and tapering at the apex, conidiogenous cells macronematous, synnematous, hyaline to plae brown, smooth, thin walled, 100-150μm long, 2-3μm broad, many celled. Conidiogenous cells phialidic, enteroblastic, monoblastic with collarette tip, each thread of the conidiogenous cell upto 3μm broad. Conidia falciform, hyaline, 10-15μm long, 2-3μm broad, smooth, thick walled with or without setulae. Setulae when present one at each end, very thin, straight or curved. The conidial mass held in mucilage appear dome shaped, covering the setae conidiogenous cells completely over growing into a cotton plug-like mass.

Material from Adilabad, Medak, Nalgonda, Khammam districts of A.P., HCIO No: 43,018. NCBL No. 57, 139. Oct. 2001. Coll. V.R. & G.S.K.

The review of the literature reveals that from India three species have been reported so far. *M. pirozynski* Varghese and Rao and *M. pleiosetosa* Vasant Rao & de Hoog, followed by *M. theobromae* Sri Valli (1995), were reported from India. However, Matsushima (1996), considered *M. pirozynski* as similar to *M. pleiosetosa*. The valid species reported from India are *M. theobromae* and *M. pleiosetosa*.

27. MONACROSPORIUM Oudemans

Ned. Krudik. Archt. Ser. 24: 236-257, 1885

Monacrosporium was introduced by Oudemans (1885), with *M. elagans* as its lectotype which was isolated from dung of rabbit from Netherlands. It produces cottony, erect, pin head like colonies. Mycelium creeping, septate, thin walled, sometimes producing ring-like traps for nematodes. Conidiophores erect, simple, branched at the tips, monoblastic, holoblastic terminal. Conidia top shaped, usually three septate, sometimes bearing micro conidia and chlamydospores.

Monacrosporium was collected from various countries like U.S.A. by Drechsler (1952), as a nematophagus fungus. Rubner (1996), described 39 species of this moniliacious fungus while revising predacious hyphomycetes. In the *Dactylella, Monacrosporium* complex, considerable work was done by many workers like Cooke (1967), Castaner (1968), Drechsler (1952), Liu and Zhang (1994), Matsushima (1971, 1975), Saxsena and Mittal (1995) and Subramanian (1963, 1971). It is known from U.S.A., U.K., Japan, China and India. *Monacrosporium* was collected from a very moist plant debris immersed in mud.

Monacrosporium eudermatum (Drechsler) Subran.J. Indian. Bot. Soc. 42: 293, 1963 (Fig. 29)

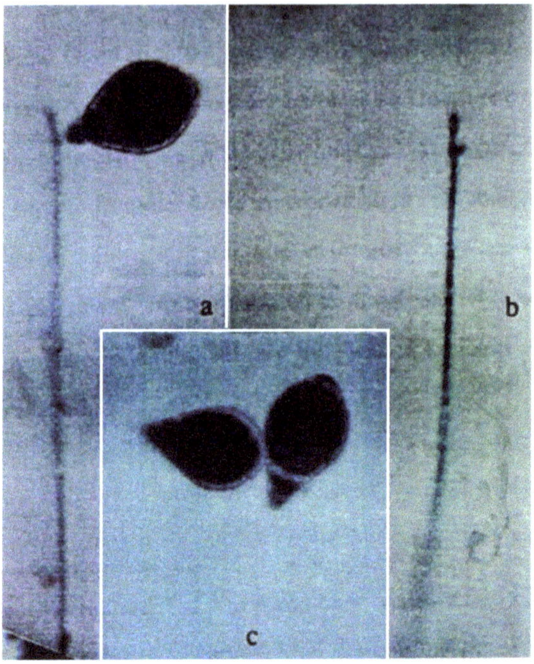

Fig. 29. *Monacrosporium eudermatum*
a,b, Conidiophores with conidiagenous cells, c, conidia (450 x)

Colonies effuse, sparse, mycelium submerged, creeping, 2-3μm broad, distantly septate, very delicate, thin walled, hyaline, sometimes forming a ring like trap to catch the nematodes.Conidiophores macronematous, mononematous, erect, straight, simple branched, 150-200μm long, smooth, hyaline 2-3μm broad. Conidiogenous cells terminal, lateral, when lateral making 90° angle to the main axis of the conidiophore short peg-like, 2-4μm broad, monoblastic. Conidia holoblastic, top shaped, 2-3 septate, first two septa at the base very near and the last septum on the top, middle cell highly inflate or swollen, 10-35μm long, 7-10μm broad in the middle, 2-3μm fine, broad at base, obtuse or conical at apex, overall shape of the conidia top shaped or spindle shaped, smooth, thin walled. Chlamydospores schizolytic and germination from any cell of the conidium.

From water bodies in Bhadrachalam, Yellandu forests of Khammam district, A.P., HCIO No: 43,019. NCBL No. 258. Nov. 2002. Coll. V.R. & G.S.K.

Monacrosporium is known to India by 5 species (Bilgrami et al. 1991). It has been isolated from *Saccarum, Munja,* soils and rotting leaves of tomato.

28. MOORELLA Raghuveer Rao & Dev Rao

Mycopathol Mycol.appl.22:50-52, 1964.

Raghuveer Rao and Dev Rao (1964), proposed *Moorella* with *M. speciosa* as its type species. It produces effuse colonies dark blackish brown, velvety. Mycelium partly superficial or immersed, septate, branched, thick walled, blackish brown in colour, 3-5μm broad. Conidiophores macronematous, mononematous, erect or straight smooth with numerous short branches formed inverticils along this stipe. Conidiogenous cells polyblastic, integrated, terminal, discrete, sub-spherical or ellipsoidal denticulate. Conidia solitary, thin walled, helicoid, septate, hyaline, ornamented or smooth.

Moorella was collected on dead wood and bark of *Aesculus* and other trees in India and Pakistan. *Moorella* remained monotypic for almost three decades. Recently, Matsushima (1993), reported *M. matiocephala* as a new species. It was collected on palm petioles in Rio aucayaco and Rio Sinchicuy Peru, South America, tributary rivers of Amazon river in Peru. This species is characteristic by the production of only one verticil on the tip of the conidiophore. The denticles are very blunt or absent. Thus *Moorella* is known by two species.

Moorella speciosa Rao & Rao, Mycopath. Mycol. Appi. 22: 50-52, 1964. (Fig. 30)

Fig. 30. *Moorella speciosa* a-d, Colony, conidiophores, conidiogenous cells, helicoid conidia (a 100 x, b-c 450 x, d 1000 x)

Colonies effuse, dark blackish brown, velvety. Mycelium partly superficial and partly immersed, septate, branched, thick walled, 3-5μm broad, septa not distantly placed. Conidiophores, macronematous, mononematous, erect, branched, in the upper region, base 3-7μm broad, apex 3-5μm broad, 250-350μm long, thick walled, smooth, branches very short, 3-4 celled, produced in verticils at intervals along the stipe, thick walled, 5-15μm long, 1-5 celled, smooth. Conidiogenous cells terminal, polyblastic, integrated on the stipe or on the branches subspherical or ellipsoidal, denticulate, 3-5μm long and upto 4μm broad, denticles 1.0μm broad, thin smooth 1-3 or more number. Conidia solitary, dry, thin walled, coilled 1-1/2 times, 4-7 septate, 10-15μm in diam., with 3-6μm filament hyaline, smooth pointed at the base and coiled, at the apex, helicoid in shape hyaline or sub-hyaline, smooth. In the present specimen the denticles proliferate further sometimes produce conidia again

On decaying wood, bark and plant debris from Papi konda, Bhadrachalam, Yellandu forests of Khammam, Nizamabad, Adilabad, Warangal, districts of A.P., NCBL No. 177, 226. Nov. 2001. Coll. V.R. & G.S.K.

Moorella remained monotypic for several years as it is a lignicolous and corticolous fungus found in all tropical forests. Bilgrami et al. (1991), reported this fungus from Nizamabad in A.P., and on the bark of *Eucalyptus* from Jabalpur, Madhya Pradesh. Vasant Rao (1973), collected it from various forests of Maharashtra and Karnataka. The new species of Matsushima is different from the type species in having strikingly only one verticil small, oval at its apex and the conidiogenous cells are monoblastic and are not polyblastic and denticulate.

29. OLPITRICHUM Atkinson

Bot. Gaz. 19: 244-245, 1894

Olpitrichum was proposed by Atkinson in 1894, with *O. carpophilum* (= *O. macrosporum* (Farlo) Sumstine, 1911), as its type species. *Olpitrichum* produces effuse, cottony colonies. Mycelium cottony thread like, anastomosing, septate, branched.Conidiophores macronematous, mononematous, erect, branched.Conidiogenous cells holoblastic, monoblastic, terminal or lateral branching characteristically in a zig-zag manner after the production of conidium. Next conidiogenous cell arises at its base making 90° which may be sometimes 1 or 2 septate.Conidia oval to spherical or sub-spherical with a minute papillae at the base. *Olpitrichum* has *Aspergillus* like anamorphic stage, which produces a conidiophore, vesicle bearing phialides and conidia in chains.

Olpitrichum is a soil borne fungus and it has also been isolated from decaying leaf litter. It is reported from European countries, Czechoslovakia and also from Canada, Australia and India.

Olpitrichum patulum (Sacc. and Berl.) Holubova-Jechova, Folia. Geo. Bot.Phytotax. Praha. 9:425-432, 1974 (Fig. 31)

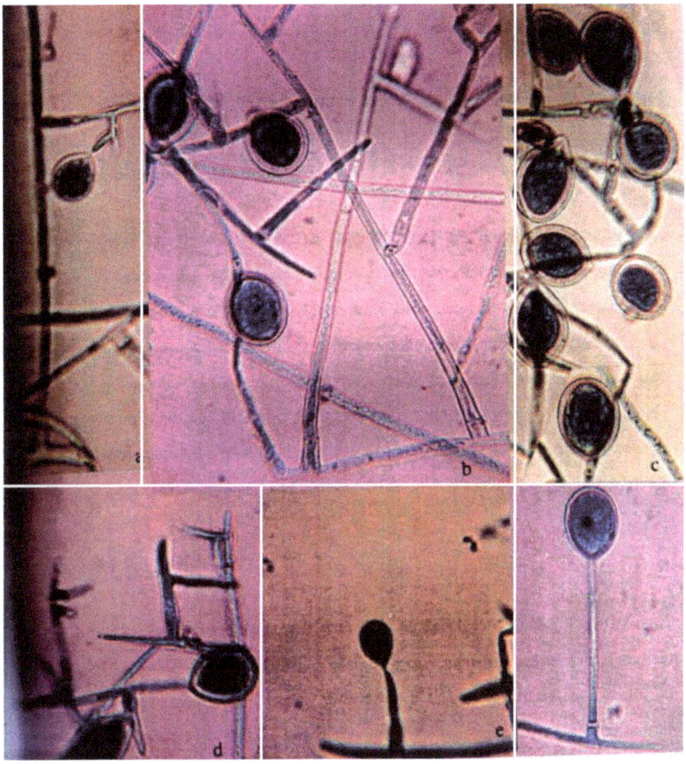

Fig. 31. *Olpitrichum patulum*
a-f, Conidiophores conidiagenous cells and conidia (450 x)

Colonies effuse, ochraceous, cottony, delicate. Mycelium superficial, anastomosing, 2-3μm broad, septate, branched, delicate, colourless, smooth. Conidiophores macronematous, mononematous, simple at base, branched at apex, 200-400μm long, 2-4μm broad, smooth, thin walled, septate, cylindrical.Conidiogenous cells terminal, lateral, characteristically making 90° to the original. Conidiogenous cells slightly narrower at the apex, cylindrical, thin walled, 1-2 celled, 10-25μm long, 2-3μm broad, monoblastic.Conidia holoblastic, terminal, solitary, ovoid to spherical with minute papilla at the base, light to pale brown, smooth 15-30μm long, 10-15μm broad, the production of the secondary conidiogenous cells is very strange. It produces *Aspergillus* like an anamorph in which conidiophores are elongated 25-50μm long, 2-4μm broad, vesicle 5-15μm in diam., thin walled, phialides 5-7μm long, 3-4μm broad, upto 1.0μm broad at the apex.Conidia 2-3μm broad or 1.5-2.5μm broad, colourless and in simple chains.

From decaying leaf litter and plant debris from Vishakhapatnam, Kakinada, Rajamundry, Bhadrachalam, Ashwaraopet forests of Khammam district, A.P., HCIO No: 43,020. NCBL No. 26, 99. Sep.2001. Coll. V.R. & G.S.K.

In fact this fungus is polymorphic in which *Olpitrichum* state and *Aspergillus* state represent synanamorphic state of the fungus.

30 . OPHIOSTOMA H. Sydow & P. Sydow

Ann.mycol. 17: 43, 1919

Ophiostoma with *O. pilifera* (Fr.) Syd., as its type species belongs to Ophiostomaceae family and Ophiostomatales order in Ascomycotina. It belongs to Sphaeriales and is very close to Xylariales. There are 4 genera and 106 species in the family Ophiostomaceae.The most common ascogenous states are *Ceratocystis, Ophiostoma* and others.Their anamorphic imperfect states are *Graphium, Leptographium, Pesotum, Verticicladiella*, and *Sporothrix.* These fungi are known to cause wilt diseases and are responsible for the loss of several economically important trees.

Ophiostoma and *Ceratocytsis*, with their 50 species cause *Dutch elm* disease, which have killed a good number of trees by causing wilt in temperate and tropical countries. *Ceratocystis* and *Ophiostoma* both produce ascomata, perithecial body rarely cleistothecial, dark bodies usually long necked ascoma with ostiolar setae, inter ascal tissues are absent and asci are not seen but they might have been formed in chains from fertile layer lining the ascomatal cavity which may be +, -, asci very thin walled, evanescent, 8 spored, ascospores hyaline, variable in shape often with eccentric wall thickening or sheath, they differ with another by their ostiolar opening and setae.The ostiolar setae in *Ophiostoma* are bulbous where as in *Ceratocystis* they are smooth. *Ceratocystis* produces *Sporothrix* anamorphic state, like *Thielaviopsis* state in addition to *Graphium* and other states. Luttrell (1951), as a matter of fact preferred *Ceratocystis* over *Ophiostoma* on the basis of priority and thought that *Ceratocystis* and *Ophiostoma* as synonymous, but in literature both names are in use. The authors prefer to use *Ophiostoma* and *Ceretocystis* separately. Since some of their collections resembles *Ophiostoma,* therefore, they prefer to describe the same as *Ophiostoma* species.

Ophiostoma species (Fig. 32)

Fig. 32. *Ophiostoma* species
a-f, Ascomata with long neck and ascospores. (a-b 100 x, c-d 450 x, e-f 1000 x)

Colonies ascomatous with spherical, elongated or flask shaped, perithecial body thick-walled, dark with long straight simple neck, opening ostiolar asci and ascogenous hyphae were not seen. Perithecial body thick walled, intra ascal body hyaline, producing evanesent asci and releasing numerous ascospores in mucilaginous mass just like oozing of the bacteria from a bacterial colony, basal ascomata 100-200μm across and 250-300μm tall with long protruding neck which is 300-500μm tall, 30-50μm wide, cylindrical made up of smooth hyphae, clubbed together, ostiolar opening funnel shaped with splashing slightly swollen at the tip, and pointed apically, sub-hyaline to brown releasing spore mass in mucilaginous surface. Ascospores elongated gattulate with 0-1 septate, smooth, sub-hyaline to pale brown, looking 2 celled. No other anamorphic state like *Graphium, Leptographium*, and others were observed. Ascospores smooth in mucilaginous mass, no germination was observed.

From Bhadrachalam, Yellandu forests of Khammam, Adilabad, Medak, Nalgonda districts of A.P., HCIO No: 43,021. NCBL No. 219, 245. Nov. 2001. Coll. V.R. & G.S.K.

31. PAATHRAMAYA Subramanian

J. Indian Bot. Soc. 35: 53-91, 1956

Subramanian (1956) proposed *Paathramaya* with *P. sudara* Subram. as it's type species. *Paathramaya* produces effused, synnematous colonies with splashing apices of the conidiophores to form a head, conidiogenous cells polyphialidic, conidia guttulose oval, 0- septate, smooth.

Paathramaya has been reported from various parts of south India by Vasant Rao (1972), Reddy (1983) and others. This fungus remained monotypic till date. In the present investigation a synnematous *Paathramaya* had been collected from Narsapur, Medak which is the type species, in addition to this, another specimen is encountered from Nalgonda district, which produces *Paathramaya* like synnemata at base individual conidiophores splash loosely. Polyphialidic conidiogenous cells producing guttulate, oval smooth conidia on distinct protuberance which is uncommon character. Therfore, it is being designated as new.

Paathramaya yadgirensis Vasant Rao et al. sp. nov. (Fig. 33)

Fig. 33. *Paathramaya yadgirensis* sp.nov. •1 Conidiophores, branches, conidiogenous cells, protuberances with conidia.

Coloniae effusae. Mycelium submersio, ex hyphs septate conidiophora macronemata, mononemata, septates, sub hyalinies vel pallide crunrei cellae conidiophore cyathiformes acropleurogene, cupulate, protuberantia Conidia monophialidica, ovata, guttulate fusce brunnea.

Typus in emortio lipro *Lantana camara* in loco Yadirguttae, Nalgonda, A.P., Indiae. NCBL No. 237. Nov. 2002. Coll. V.R. & G.S.K.

Colonies effuse in scattered loose clusters, mycelium suburged septate, smooth, pale to light brown, 3-5μm wide. Conidiophores in loose clusters not synnematous, macronematous, mononematous, erect, branched, smooth, pale to light brown, 150–300μm long, 3-7μm broad, septate, thick walled at base white thin alive at the apex. The conidiophores part apices of the conidiophores. Conidiogenous cells polyphialidic acropleurogenous 10-15μm tall, 3-7μm wide at the widest, the cupule producing prominent protuberent stalk like structure which bears conidia. Conidia phialidic, arising from the cupule on protuberent stalk like structure which collapses while dispersing the conidium, conidia 0-septate, 1- celled, smooth, oval, thick walled, 10-15 x 3-7μm, conida secession rhexolytic.

Collected on dead *Lantana camara* stalks Yadgirigutta, Nalgonda district, A.P., NCBL No. 237. Nov. 2002. Coll. V.R. & G.S.K.

This is evident that *Paathramaya yadgirensis* differs from the type species i.e. *P. sundara* Subram., in not having synnematous colonies and having branched conidiophores in addition sharing the generic characters of conidigenous cells and conidia. Therefore, it is being described as new. The specific epithet denotes the place of its collection viz., Yadagirigutta.

32. PERICONIA Tode ex Persoon

Syst. Meth. Fungo., 686, 1801

Periconia was created by Tode with type species as *P. lichenoides* Tode ex Mirat, which is lost, hence *P. byssoides* Pers. is established as lectotype.

Periconia is recognised by its effuse, small and compact grey or brownish black, stromatic colonies. Conidiophores always mononematous (only one species with synnema), macronematous or micronematous variable in shape and size. Conidiogenous cells mono or polyblastic, determinate or indeterminate, termninal or intercalary, discrete, variable in shape. Conidia holoblastic, zero-septate, in catenate, simple or often branched chains, spherical or variable in shape, acropetal in succession but mature from apex to base. *Periconia* is one of the most studied conidial fungus which is cosmopolitan in distribution.

Periconia is an extremely common form genus found on dead herbaceous stems and decaying leaves and cause leaf-spots. It is also associated with other fungi like *Corynespora cassicola, Epicoccum purpurascens* and other as a saprophyte or sometimes parasitic. It is cosmopolitan in distribution. It occurs in U.K., U.S.A., Canada, British, Solomon Islands, India, Zambia, Australia, New Zealand, Sudan, Ethiopia, Brazil, Cuba, Kenya, New Guinea, Hongkong, Iraq, Israel etc. It occurs on a number of angiospermic host plants in forest field, agricultural crop plants like jawar, *wheat, Chenopodium* and grasses. It has been isolated from soil as a rhizosphere fungus and isolated from seed surface and air.

Subramanian (1955), extensively studied the species of *Periconia* found in India. Inspired by his work Raghuveer Rao and Dev Rao (1964 a & b), published a number of new species and presented a beautiful illustrated account and traced its evolutionary trends in colony, conidiophores, conidiogenous cells, and conidial head etc. From these the scope for further studies in India, was taken up by Satyanarayana (1983). India is a very rich country in the distribution of species of *Periconia.* So far from different parts of India more than 50 species have been proposed by different workers (cf. Bilgrami et al.1991).

1. *Periconia atropurpurea* (Berk. & Curt.) Litvnov, Opredelitel Mikroskopieuskikh Pourennykh Gribov, Moniliales, Aspergillaceae, 147, 1967. (Fig. 34)

Fig. 34. *Periconia atropurpurea* a. Colony (100 x), b-c, Conidiophores and conidia (450 x).

Colonies pulvinate, dark chocolate to black often staining the substratum reddish purple 0.5-1*mm* across. Mycelium immersed in substratum, visible hyphae 2-3μm wide, septate, branched, cylindrical, smooth or thick walled sometimes a weak stromatic body of pseudoparenchymatous cell is formed below the epidermis and disrupting it conidiophores come out. Conidiophores macronenmatotus, mononematotus, 150-270μm long, 2-4μm broad at base, 5-7μm broad at apex, upper part sterile, shorter than the fertile part, often curved or bent at the tip, pale to dark wall, smooth. Conidiogenous cells distributed in the fertile part. 1– many celled cylindrical, polyblastic, 5-20μm long, 3-5μm broad, smooth, plae to dark brown lateral or intercalary. Conidia holoblastic in acropetal simple or branched chains but maturing from apex to base, spherical, dark reddish brown, verrucose or finely echinulate 7-10μm in diam. In the chains they held together by very minute isthumii which collapse in due course. No conidial secession, germ pore or germination of conidia was observed.

On decaying leaves of *Borassus flabellifera* from Nizamabad, Parkhal, Nirmal, Bhadrachalam, Yellandu, Ashwaraopet forests of Khammam district, A.P., HCIO No: 43,022. NCBL No. 88, 150, 211. Oct./Nov. 2001. Coll. V.R. & G.S.K.

2. *Periconia byssoides* Pers. ex Schwein. Synopsis Fungorum. Carolinae superiors. P. 125, 1822. (Fig. 35)

showing apex with septum conidiogenous cells and conidia (a-b 450 x, c 1000 x)

Colonies black, hairy, 1-3 *cm* across, each conidiophore under stereobinocular microscope looks like round pin headed structure, black in colour. Mycelium submerged in the substratum, visible hyphae hyaline to pale brown smooth, branched, septate, 1-5μm in diam., stromata partly inserted in the substratum flattened or inverted saucer shaped, dark brown, 50-80μm in diam.,10-25μm high.Conidiophores macronematous, mononematous arising from the flexuous septate, black shining by reflected light, smooth 200-1000 or more millimicron tall,10-20μm broad at base, 5-15μm broad below the head.Conidiophores always simple and branched the branching of the stipe is seen only when the apex is injured. The apical part of the stipe bearing spherical or sub-spherical, thin or thick walled, polyblastic conidiogenous cells.The stipe with a dome shaped apical cell cut off by a septum, conidiogenous cells formed over the apex and in a ring below the septum.Conidiogenous cells attached to the stipe and secondary conidiogenous cells may arise from the primary conidiogenous cells bearing short chains of conidia. The chains of conidia constitute the conidial head which is 30-110μm across, black in colour.Conidia spherical, brown, verrucose at first and later on evenly thickened 10-15μm in diam. Conidia produced in acroptal chains but maturing from apex to base. In the chains isthmii could not be observed.The conidia germinated readily in tap water producing one or two germ tubes.

On decaying twig and leaf litter from Medak, Nalgonda, Adilabad, Bhadrachalam, Yellandu, Ashwaraopet forests of Khammam district, A.P., HCIO No: 43,023. NCBL No. 90, 121. Nov. 2001. Coll. V.R. & G.S.K.

P. byssoides was reported earlier on stems of various plant materials from Chingalpat, Nilagiris in Tamilnadu and dead stems of unknown leguminous plant near Narsapur forest in Medak district of Andhra Pradesh. It is common in other states of India and other countries like, U.K. West Africa, Sudan. Malaya, Australia, New Zealand, West Indies and U.S.A.. This fungus is comparable to *P. cookei, P. shamala* and *P. jabalpurensis*.

3. *Periconia cookei* Mason and Ellis, Mycol. Pap. 56: 57-58,1951.

(Fig. 36)

Fig. 36. *Periconia cookei* - a-d, Conidiophores showing apex not septate conidia all along the head (a-c 450 x, d 1000 x)

Colonies effuse hairy pin head like, umber to brown in colour. Mycelium superficial, creeping septate, branched, unevenly thickened, 3-5μm broad, stromatic body is/of weak cells unequal in size. Giving rise to one or two conidiophores. Conidiophores erect, macronematous, mononematous, simple, unbranched, septate, cylindrical, 1000-1200μm long, 8-20μm broad at base, 10-25μm broad at apex below the conidial head, stipe without apical cell, conidigenous cells polyblastic in primary or secondary order, variable in shape, thin walled, producing chains of the conidia and formings a black, spherical conidial head about 150μm broad. Conidia spherical, pale to mid brown, thin walled, minutely and distinctly echinulate 9-15μm in diam., in acropetal chains. The conidial secession and germination were not observed.

On unidentified twigs and leaf litter, from Adilabad, Nizamabad, Nalgonda, Bhadrachalam, Yellandu, Ashwaraopet forests of Khammam district., A.P., HCIO No:43,024. NCBL No. 42, 181. Nov. 2001. Coll. V.R. & G.S.K.

Periconia is very interesting genus. Its colonies are mimicry of some angiospermic plants. A view of the colony, under the binocular microscope reveals interesting features. Ingram and Hudson (1995), have presented a beautiful account on form and shape of plants and fungi. It takes into different aspect of morphological study reminding an aerial view of land scape, canopy of different kinds of plants. Due to certain limitations the authors are restricted to include only three species of this beautiful genus from Khammam district.

Key to the species of *Periconia*

I. Conidiophores macronematous, micronematous, mononematous or synnematous. Conidiogenous cells polybiastic. Conidia in acropetal chains, maturing from apex to base..*Periconia*

1. Conidiophores macronematous. Conidial head indistinct ....*P.atropurpurea*
2. Conidiophores macronematous, apex septate, unbranched producing conidia below the septum and above to form head...*P.byssoides*
3. Conidiophores macronematous, apex not septate. Conidia borne all along the head..*P.cookei*

33. PERICONIELLA Saccardo

in Saccardo & Berlese, Atti.ist.Veriete. Sci.Ser. 6, 3: 727, 1885

Periconiella was created by Saccardo (1885), with *Periconiella velutina* (Wint.) Sacc., as its type species. *Periconiella* occasionally produces compact but usually effused, hairy or velvety colonies. Mycelium superficial or immersed, hyphae septate, branched. Conidiophores macronematous, mononematous or synnematotus branched.Conidiogenous cells polyblastic, sympodic, cicatrised with prominent conidial scars. Conidia holoblastic, solitary, rarely in chains 0-few transversely septate, brown, smooth or verrucose.

Periconiella occurs on variety of substrates as a parasite and saprophyte, it was reported from several African countries, S.America, New Zealand, Australia, Papua New Guinea, Malaya, Sarawaka, India and other countries. This fungus creates difficulty in distinguishing it from *Cladosporium, Ramichloridium, Cercosporidium, Cercospora* and others. However, production of verticillate or tuft of branches at the tip of conidiophore characteristic scars at the base of the conidia and conidiogenous cells and absence of strormatic body are main outstanding characters.

Periconiella cocoes Ellis, Mycol. Pap. No. 111: 23, 1967. (Fig. 37)

Fig. 37. *Periconiella cocoes*
a-d, Synemata, conidiogeunous cells and conidia (a 100 x, b-d 450 x)

Colonies effuse, irregular, zonate, dark brown, velvety, synnematous (Coremia,), loosely arranged mycelium partly superficial and partly immersed, passing through the stromata, never forming stromatic body, 2.8-5.6*μm* broad, olivaceous brown to brown.Conidiophores macronematous, synnemetous rarely, mononematous, simple or branched, in fascicles, terminaly and laterally, the synnemata (Coremium) forming distinct stipe and a complex conidial head. The synnema may be branched in fascicles, pale brown or brown, smooth, 400-700*μm* long and 40-70*μm* broad. The conidiophore branched repeatedly to produce primary or secondary or tertiary branches. Each branch, a conidiogenous cell, cylindrical, pale brown, variable in length 2.8-4.2*μm* thick, polyblastic, cicatrised with numerous acropleurogenous prominent scars, conidia holoblastic, solitary, obclavate or shortly clavate, pale to mid dark brown, smooth, 2-celled, slightly constricted at the spetum, upper cell thicker, sub-globose, slightly larger and lower cell narrower, paler, smaller than upper cell, 4.2-8.4*μm* long, 2.8-4.2*μm* broad. Conidial secession schizolytic and conidial germenation through corner cell.

On decaying leaves *(Cocus nucifera* and *Borassus flabellifera)* from Nalgonda, Warangal, Bhadrachalam, Kottagudem forests of Khammam district, A.P., HCIO No:43,025. NCBL No. 33, 128, Oct./Nov. 2001. Coll. V.R. & G.S.K.

Periconiella cocoes Ellis is new record to Indian Mycoflora (Bilgrami et al. 1981), Review of Indian literature reveals that *Periconiellia* has been reported from India as a parasite causing leaf spots on various wild and cultivated crop plants. But this species has close affinity in just conidial form with *Spadicoides bina* (Corda) Hughes, and *Diplococcium* state of *Helminthosphaeria clarariarum* (Tule) Fuckel.

34. PESTALOTIA De Not

Mem. R. Acad. Sci. Torino 2,3: 80, 1839

De Not (1839), introduced the genus *Pestalotia* with *Pestalotia pezizoides* De Not, as its type species. *Pestalotia* is a parasitic pycnidial coelomycetes causing leaf spot diseases in many agricultural, horticultural and forest crop plants. Its conidiomata are acervular or pycnidial, circular solitary with or without hallow around the dark region, with or without operculum. The pycnidium or acervulus is made up mycellium, septate, hyaline to brown, pseudoparenchymatous and producing conidiophores in the cavity of pycnidium or free end of the acervulus, macronematous on mat of conidiomata, long and branched conidiogenous cells 1-3 annellate, endogenous, conidia 5, distoseptate, 4 median cells brown and end cells hyaline, apical cells or basal cell brown, and cells hyaline apical cells or basal cell with simple or branched appendages.

Pestalotia is very common in A.P., causing leaf spot disease on *Mango*. *Pestalotia* has wide range of hosts of economically important crops. There are about 200 species in this genus mostly host specific. Bilgrami et al. (1991), recorded 95 species in India. A good amount of work has been done on this taxon resulting in recognition of several allied genera and species. However, Sutton (1971), has recognised only one species i.e. *Pestalotia pezizoides* De Not. Thus according to Sutton(Op.cit.), *Pestalotia* is monotypic.

Pestalotia pezizoides De Not, Mem.R. Acad. Sci .Torino. 2, 3 : 80, 1839. (Fig. 38)

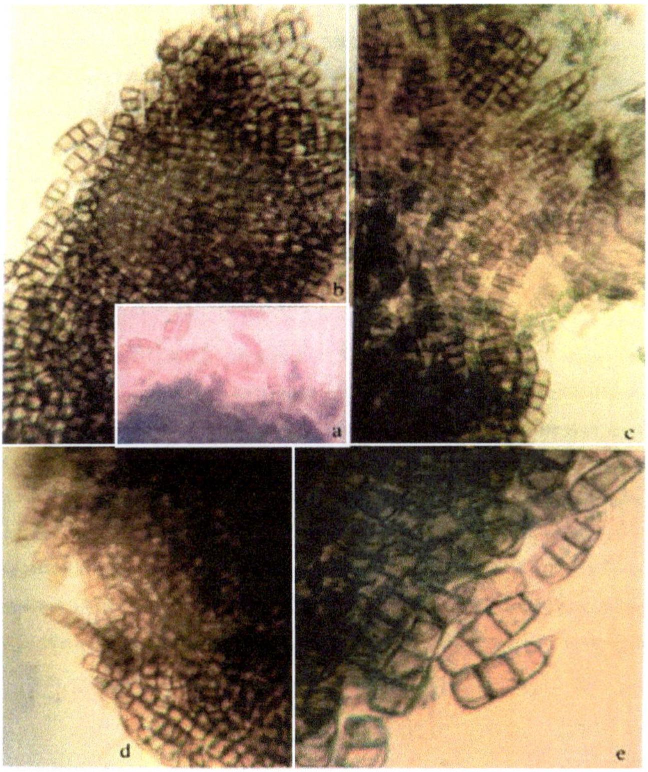

Fig. 38. *Pestalotia pezizoides* - a, Colony (100 x), b-e, Conidiomata with conidial mass (450 x, 1000 x).

Infection spots, round, black to dark brown, conidiomata, sub-epidermal complex. Mycelium immersed, incorporated to form a mat of pseudoparenchymatous, the basal part attached to host tissue dark brown to black fertile part bearing sub-hyaline cells consisting conidiophores. Conidiophores macronematous, mononematous, erect, simple 2-3 celled, 2.8-5.6μm broad variable in length. Conidiogenous cells percurrent, annellidic 2-3 annellations an each conidiogenous cell 2.8-4μm tall. Conidia holoblastic, endogenous, terminal pharagmoconidia. The basal and apical cell hyaline to sub-hyaline pointed sometimes giving rise 2-3 fine hyaline setulate 14-56μm long and 3.5-5.6μm broad. The conidial mass oozes out of the operculum through sooty structure. Conidial secession not observed, conidial germination through base and apex.

On living and dead *Mango* leaves from Adilabad, Vishakapatnam, Medak, Warangal, Ashwaraopet and Bhadrachalam forests of Khammam district, A.P., HCIO.No: 43, 026. NCBL No. 66, 195, Nov. 2001. Coll. V.R.& G.S.K.

Pestalotia is a controversial taxon known with its anamorphic and teleomarphic nature, due to very wide range of host specificity and less morphological identity, it is considered as monotypic. It is cosmopolitan in distribution.

35. PHAEOISARIA Hoehnel

Sber. Akad. Wiss. Wien Abt., 1: 118, 329-330, 1909

Hoehnel (1990), proposed *Phaeoisaria* with *P. bambusae* Hoehnel, as its type species but Hughes (1958), while studying the European herbaria observed that *Phaeoisaria clematidis* (Fuckel) Hughes, was already known and synonymised Hoehnel's specimen. *Phaeoisaria* is characterised by effuse, hairy, prominent, synnematous colonies covered with white conidial mass. Mycelium immersed. Conidiophores macronematous, synnematous individual threads branched towards their apices splashing out at apex, upper half or 2/3 portion fertile in each synnema. Conidiogenous cells polyblastic, sympodial, clavate, cylindrical, discrete, denticulate. Conidia dry, acropleurogenous, fusiform, ellipsoidal or sub-spherical, hyaline or sub-hyaline, smooth, zero-septate.

Its widely distributed and common conidial fungus found in almost all parts of world including tropic, subtropical or temperate zones. Most of the above mentioned workers like Ellis (1971, 1976), Seifert (1985), and others have discussed its morphology and taxonomy. In the present work few samples are treated under *P. clematidis* (Fuckel) Hughes, and *P. caffera* Matsushima.

1. *Phaeoisaria clematidis* (Fuckel) Hughes, Can. J. Bot. 36: 795, 1958. (Fig. 39)

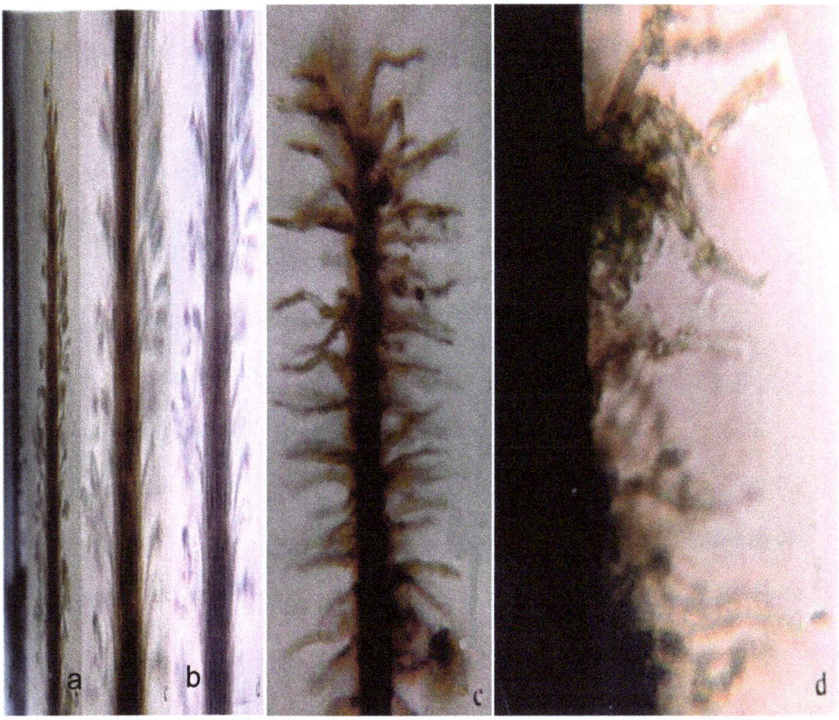

Fig. 39. *Phaeoisaria clematidis*
a-d, Synnemata, conidiogenous cells with conidia (100 x, 450 x, 1000 x)

Colonies effuse, black hairy, under hand lens or binocular microscope upper half or upper two third position covered with a white or pale grey powdery mass of conidia (when collected in rainy season otherwise not). Mycelium immersed, visable hyphae, septate, sub-hyaline to pale brown, 2-3μm wide, branched, stroma visible, poorly developed. Conidiophores macronematous, synnematous, textura angularis parallel, cylindrical, septate, splashing in upper region along the sides 200-500μm long, 10-30μm thick at base and 25-40μm in the conidial region, individual threads septate, sub-hyaline, smooth

2-3μm thick, conidiogenous cells, polyblastic, integrated, terminal, discrete, sympodial, subulate, cylindrical or clavate, tapering to fine point denticles which may be upto 7 per conidiogenous cells, sub-hyaline to pale brown. Conidia solitary dry, acropleurogenous, simple, fusiform or ellipsoidal, hyaline to sub-hyaline, 3-9μm long and 1-2μm broad. Conidial secession and conidial germination could not be observed due to minute denticular nature and small size.

On fallen branches, leaves of many unidentified plants from Nizamabad, Adilabad, Kakinada, Sri Kakulam, Bhadrachalam, Yellandu, Ashwaraopet of A.P., HCIO No: 43,027. NCBL No. 181, 288 Nov. 2001. Coll. V.R.& G.S.K.

2. *Phaeoisaria caffera* Matsushima, Matsush.Mycol.Mem.No.9:19, 1996. (Fig. 40)

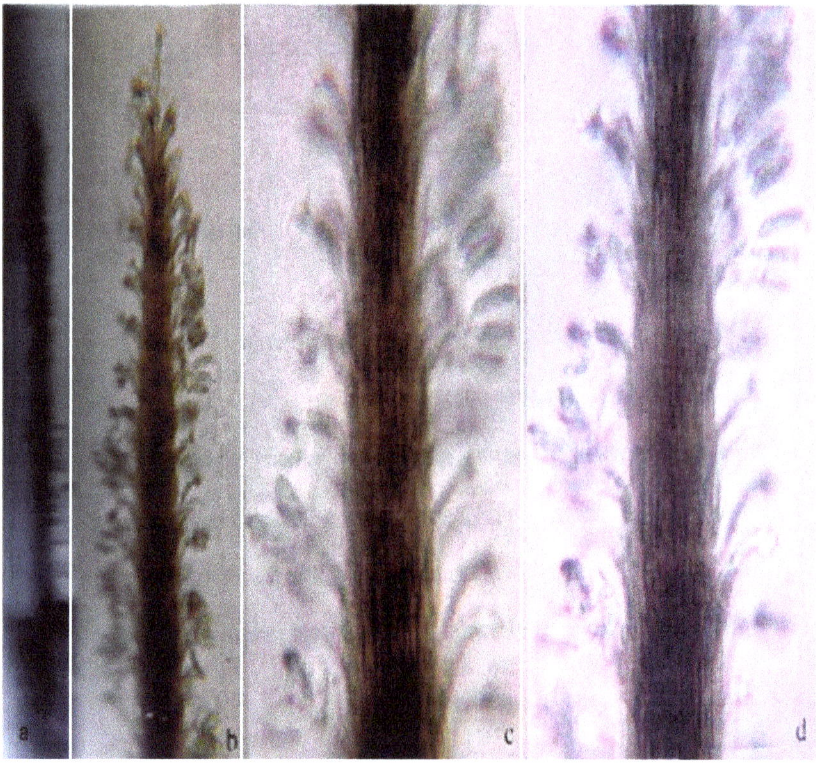

Fig. 40. *Phaeoisaria caffera* - a, Synnemata (100 x), b, apical portion of synnemata (450 x), c-d, conidiogenous cells with conidia (1000 x).

Colonies effuse, tapering columnar, synnematous hairy whitish catkin like when in full with conidial production, solitary or in tufts of 2-3. Mycelium septate, immersed, sub-hyaline to pale brown 2-3μm wide, branched stroma poorly developed. Conidiophores macronematous, synnematous, textura angularis parallel, cylindrical septate, splashing in upper region along the sides 200-500μm long, 10-30μm thick at base, 25-40μm broad in the conidiogenous cells region. Individual threads septate sub-hyaline, smooth 2-3μm thick. Conidiogenous cells polyblastic, integrated, terminal, discrete, sympodial, subulate, tapering or fine pointed bearing conidia. Conidia holoblastic, thin walled 1-celled, solitary, obovate, 7-12μm long and 2-3μm broad smooth whitish or greyish brown in mass.

On decaying plant debris and twigs from Karimnagar, Bhadrachalam, Yellandu forests of Khammam district, A.P., HCIO No: 43,028. NCBL No. 234, 314. Nov. 2001. Coll. V.R.& G.S.K.

Phaeoisaria caffera is very close to *P. clematidis* it differs only in conidial dimensions. Matsushima (1996), compared his species with *P. clematidis* described by Deighton (1974), Sutton (1973), Romero (1983), and Chang (1989), and concluded that his species had larger conidia. The species of Khammam agrees with *P. caffera* in all respects i.e. in having larger conidia than *P. clematidis*. Hence it is a new record to Indian mycoflora.

Phaeoisaria specimens show whitish conidial mass only when they are collected in humid weather or examined the specimens kept in moist chamber. Its cosmopolitan occurrence suggests that it degrades cellulose, hemicellulose and lignin.

Key to the species of *Phaeoisaria*

I. Conidiophores macronematous, synnematous. Conidiogenous cells denticulate, polyblastic splashing...*Phaeoisaria*

1. Conidiogenous cells polyblastic, denticulate, elongated 1-celled. Conidia smaller than 7μm...............*P.clematidis*
2. Conidiogenous cells polyblastic, denticulate, elongated 1-celled. Conidia 12 x 2-3μm ....*P. caffera*

36. PHALLUS Junius ex Linnaeus 1753

Phallus was introduced by Junius ex Linnaeus in 1753, with *P. impudicus* as its type species. *Phallus* is known by its basidiocarps which are egg like about 5 *cms* diam., which in turn develop on extentive whitish mycelial cords to bear the honey comb like receptacle which bears the fertile part of the gleba, within the young gleba are cavities lined by basidia which bear upto 9 spores. They are liberated on disintergration of basidia and other fleshy part. The stalk of the *Phallus* measures between 5-15 *cms.* The section of the mycelial cord show thin papery outer and inner peridium wide mass of jelly making the middle peridium central fruit body is hollow stipe ending up in a folded honey comb like receptacle.

There are 18 species found in the world and usually are known as stink horns, Bilgrami et al. (1991), reported 5 species from India. In the present studies *Phallus* was collected on the soil in egg like stage, as basidiocarp.

Phallus impudicus ex Webster, Introduction to Fungi, Pp.480-483, 1980. (Fig. 41)

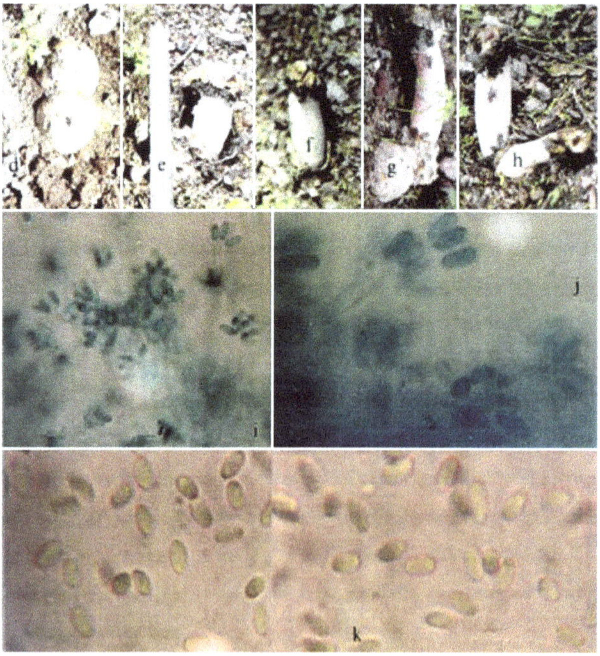

Fig. 41. *Phallus impudicus* - a-b Developing basidiocarp, C V.S. of basidiocarp showing development, e-h, Developing stipes with caps, i-k basidia with basidiospores, (i 450 x, j-k 1000 x)

Phallus colonies or basidiocarps were found egg like, about 5 *cms* in diam. Mycelium immersed in the soil forming cord like structure consisting of septate, pale to brown, thin walled hyphae, with clamp connections, sometimes skeletal cells, rather thick walled cells were also seen. Basidiocarps differentiated in the longitudinal section into outer wall i.e. peridium, inner peridium, middle peridium made up of wider mass of jelly like substance. In the fertile undeveloped part sections were taken to study the basidia which are elongated, each basidium produces 4-9 basidiospores which are elongated bullet like, brown coloured 2-3μm broad, 4-6μm long smooth.

From near by water bodies in Bhadrachalam, Kottagudem, Adilabad, Srikakulam districts, A.P., NCBL No. 230, 305. Nov./Dec. 2002. Coll. V.R. & G.S.K.

37. PLEUROTUS (Fries) Kummer

ex Saccardo, Syll. Fung. 5: 348-370, 1887.

Pleurotus is a nomen - conservatum, it belongs to Lantinacceae or Tricholomataceae family of Agaricales order of Basidiomycotina. In Agaricales 15-16 families are recognised. It is known by 50 species especially growing on decaying wood. This is cosmopolitan in distribution. *P. ostreatus* is the common Oystercarp which damages wood, besides being edible and commercially cultivated. But its allied species viz., *P. olearius* is deadly poisonous mushroom. Therefore, one should take utmost care in handling and processing the macrofungi. *Pleurotus* does not produce a distinct stalk the stipe which is funnel shaped and its shape is like a shell like layer. The pileus is oyster or shell shaped in *Pleurotus ostreatus*. The young cap is deeply, bluish but soon they turn pale brown, the surface is smooth and moist with age turned down words. The stipe is white, if present it is attached laterally, but usually it is absent. From Khammam forest a *Pleurotus* like fungus was collected. Recently, Peterson and Hughes (1997), have proposed a new species of this genus.

Pleurotus ostreatus (Fries), Kummer ex Soothill and Fairhourst, The New Field Guide to Fungi, 1978. (Fig. 42)

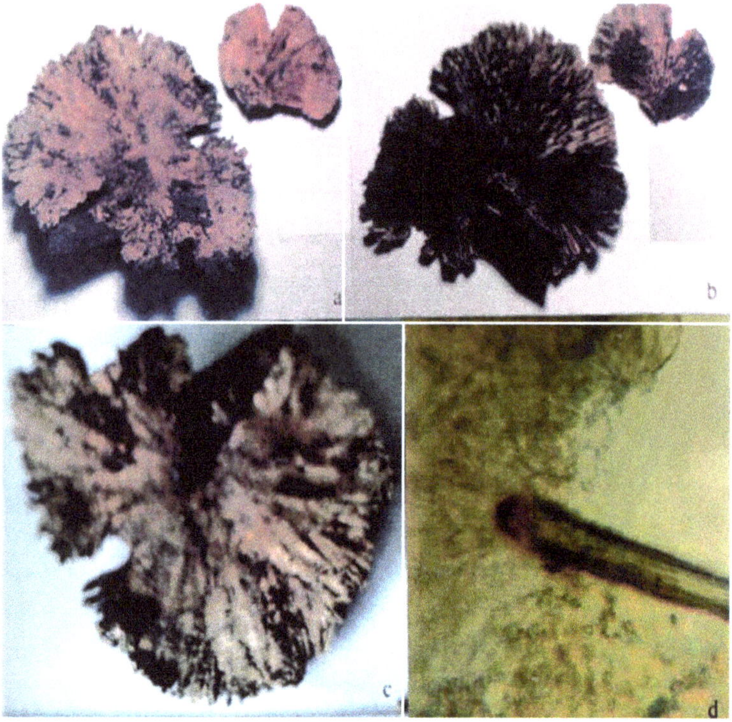

Fig. 42. *Pleurotus ostreatus*
a-c, Basidiocarp, d-e, Skeletal cells and basidiospores (450 x)

Fungus oyster like, capitate cap or pileus 3-15*µm,* shell shaped with smooth, with incurved margin, buff to pale cinnamon coloured on drying, flabelliform with attenuated base, margin in sized straight, finally veil absent. Surface smooth on apicutaneous made up of repent parallel hyphae, 5-12*µm* broad trama made up of 10-15*µm* broad hyphae, lamellae decurrent with slight inter connections towards the base unequal, with sets of lamellulae cream coloured becoming coriaceous fleshy moderatly crowded margin entire, trama homoisomerous irregular 5-8*µm* broad surface euhyniemal, sub-hymenial 10-30*µm* thick pleurocystidia absent, edge sterile with clavate cheilocystidia sometimes constricted with broad obtuse apex, hyaline, thin walled, 18-20*µm* long, 4-6*µm* broad, basidia cylindrical, tetrasporic, sterigmata 1-1.5*µm* long, collapsible basidia hyaline, 15-20*µm* long, 3-5*µm* broad, smooth basidiospores, 5-10*µm* long, 2-4*µm* broad, cylindrical, hyaline, smooth.

From around water bodies in Warangal, Adilabad, Bhadrachalam, Yellandu forests of Khammam district, A.P.. HCIO No: 43,050. NCBL No. 160, 199. Nov 2001. Coll. V.R.& G.S.K.

Pleurotus is an edible fungus and it is cultiviable but the high temperature in summer and the ecological conditions of the region whether they support mushroom cultivation is not known. It would be pertinent to state that Khammam forest is also a viable area for cultivation and marketing of this fungus.

38. PITHOMYCES Berkeley & Broome

J. Linn. Soc. London 14: 100, 1873

Pithomyces was proposed by Berkeley and Broome in 1873, with *P. flavus* as its type species. *Pithomyces,* produces effuse, black granular colonies. Mycelium superficial, creeping, septate, branched. Conidiophores semi-macronemtous, peg-like, erect or bent, one celled, cylindrical, smooth or verrucose. Conidiogenous cells terminal, monoblastic. Conidia holoblastic, phragmo or dictyo or one celled, conidia with variable ornamentation. Dehisence of the conidia by fracturing the conidiogenus cells and the conidia carries it as a small frill.

Pithomyces was redescribed and recircumscribed with clear modern generic concepts by Ellis (1960), and giving the clear illustrations of *P.flavus, P.chartarum, P.maydicus* and other species. Ellis (1971), described number of species collected in Common Wealth Countries including India. From India 9 species have been reported on decaying leaf litter of various plants like *Chenopodium, Typha, Flambago, Cassia, Cajanus, Cannabis, Heteropogon, Symbopogon, tectona, Partheinium, Lantana, Eucalyptus, Coix, Vitis, Vigna,* and leaf of *Arachis, Borassus* and decaying paper, card board, plywood and others. The species of *Pithomyces* have been isolated from mud, soil and air. It has been reported from U.S.A., Australia, and other countries. *Pithomyces* is basically a tropical or subtropical hyphomycete. It is also present in the temperate countries. In Australia *Pithomyces* is responsible for facial eczema in sheep and cattle. It produces a toxic substance known as sporidesmin and sporodemolites. These chemicals are harmful to human beings according to Brewer et al. (1989). Roux (1986), attempted its teleomorph which is ascogenous in nature. In the search of literature, it has been found out that *P.ellisii* Vasant Rao and Chary is a mycoparasite on *Excipularia, Sporidesmium, Cladosporium* and others.

Pithomyces chartarum (Berk. and Curt.)Ellis,Mycol.Pap.No.76:1-19, 1960 (Fig. 43)

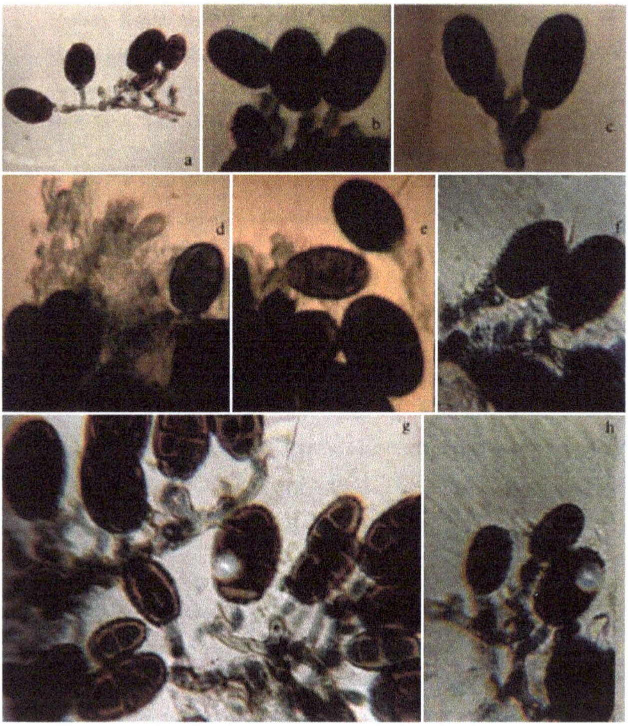

Fig. 43. *Pithomyces chartarum*
a-h, Colonies, conidiogenous cells and conidia (450 x, 1000 x)

Colonies effuse, black, produces granular, spreading about 1 *cm* across. Mycelium creeping, anastomosing, sub-hyaline to pale brown, septate, branched, 3-4μm wide smooth or verrucose. Conidiophores 0-1-celled, cylindrical, erect or flexuous, 3-10μm tall, 2-13μm broad, sub-hyaline to pale brown smooth or verrucose.Conidiogenous cells terminal, indiscrete, monoblastic, usually fracturing at the time of liberation of conidia.Conidia holoblastic 3-4 transversely septate and 2-3 longitudinally septate, muriform, oval or oboval, cylindrical, clavate shaped with pale to dark brown, smooth or verrucose ornamentation, 15-30μm long, 10-15μm broad. Sometimes its morphology is greatly influenced by temperature and nature of the substratum. Conidial secession rhexolytically and carrying the part of the conidiogenous cell as frill along with them. Conidial germination through any cell of the conidium.

From plant debris, decaying leaf litter from Bhadrachalam, Yellandu. Ashwaraopet forests of Khammam, Adilabad, Medak districts, A.P., HCIO No: 43,029. NCBL No. 173, 253. Nov 2001. Coll. V.R.& G.S.K.

Pithomyces chartarum is collected in all seasons. Its pigmentation and oramentation are influenced by temperature and nutrients in the substratum. The specimens collected in colder parts of the year are light brown in colour,where as those collected in hotter periods of the year are more pigmented, dark, thick walled highly ornamented. Its biology in India is not properly studied and practically there is no record of this fungus as pathogen for sheep and cattles in India.

39. PSEUDOBOTRYTIS Krzemien & Badura

Acata Soc. Bot. Poloniae 23: 174, 1954

Pseudobotrytis was proposed by Krzemien and Badura (1954), from Poland with *P.terristris* (Timonin) Subramanian, as its type species. It produces effuse colonies, characteristic of small pin head like open umbrella bars. Mycelium superficial or creeping. Conidiophores erect, simple, branched only at apex to form all umbell like structure, conidiogenous cells tips of umbell polyblastic, denticulate. Conidia one or two celled whitish coloured or brown.

Pseudobotrytis though reported from Poland but it was also known from U.K. Canada, Japan. From India two species are known from Rajastan and Tamilnadu, according (Bilgrami et al.1991). Matsushima reported *P.bisbyi* and *P.terristris* (Tim.) Subram., from Japan.

Pseudobotrytis bisbyi Timonin, Ceiba 9: 28, 1961. (Fig. 44)

Fig. 44. *Pseudobotrytis bisbyi*
a-h, Conidophores with conidiogenous cells with denticles and conidia (450 x, 1000 x)

Colonies effuse, pin head like, whitish brown or black, solitary or in tufts, of conidiophores. Mycelium superficial or immersed, sub-hyaline to pale brown, thin or thick walled, septate, branched. Conidiophores arising laterally on the creeping hyphae, macronematous, mononematous, cylindrical, septate, smooth, thick walled, pale to dark brown, 150-300μm long, 3-7μm broad at base, 2-3μm broad at apex, branches confined to the tip of the conidiophore. The conidiogenous cells polyblastic, terminal forming an umbrella, straight or curved granulate. Conidiogenous cells polyblastic, denticulate, proliferating, 10-25μm long, 2-4μm broad and 3-6μm broad at the conidiophorous zone, denticle 1-2μm long. Conidia oblong, ellipsoid, continuous, smooth, brown, 3-6μm long, 2-3μm broad, conidial secession schizolytic.

From decaying leaf litter and plant debris from Medak, Sri Kakulam, Bhadrachalam, Yellandu forests of Khammam district, A.P., HCIO No: 43,030. NCBL No. 186, 261. Nov./Dec. 2001. Coll. V.R.& G.S.K.

Pseudobotrytis is collected as a saprophytic on leaf litter as decomposing fungus. The specimen of Khammam is slightly variable as far as the conidiogenous cells are concerned. The proliferation and denticles are prolific.

40. RAMICHLORIDIUM Stahal ex de Hoog

Stud. Mycol. 15: 59-86, 1977

Ramichloridium Stahel (1937), was validated by providing Latin diagnosis by de Hoog in 1977, who considered 14 species and varieties under this name. *Ramichloridium* was divided into 2 sections by de Hoog, as *Ramichloridium* section *Ramichloridium*, and *Ramichloridium* section *pallesenes*. *Ramichloridium* takes in fungi producing rapidly growing furinose to lanose brown or olivaceous green colonies. Mycelium submerged thin walled, hyaline, smooth, conidiophore from aeral or submerged hyphae, macronematous, mononematous, branched or unbranched. Conidiogenous cells terminal or lateral, denticulate, rachis swollen, polyblastic producing hyaline to sub-hyaline, one celled smooth or verrucose, subglobose or fusiform. Conidia of the type species of this genus were selected of *Chloridium apiculatum* Miller et al. culture.

Ramichloridium reminds several genera like *Veronia*, *Chloridium*, *Pleurographium*, *Rhinocladiella* and other genera. From India *Ramichloridium* earlier was reported under different names like *Veronia, Chloridium* and others. It was Vasant Rao and de Hoog (1986), who reported a new species namely *Ramichloridium faciculatum* growing on decaying wood from Thirathahalli, Karnataka, with a synnanamorph of *Exophiala*. Subramanian (1971) reported this fungus as *Chloridium indicum* which is now *Ramichloridium musae* (Stahel ex Ellis) de Hoog. *Ramichloridium* was collected from various countries like Africa, S. America, India, Japan, and others, on natural substrate, like soil and others.

Ramichloridium musae (Stahel ex Ellis) de Hoog, Stud.Mycol.15: 62-63.,1977. (Fig. 45)

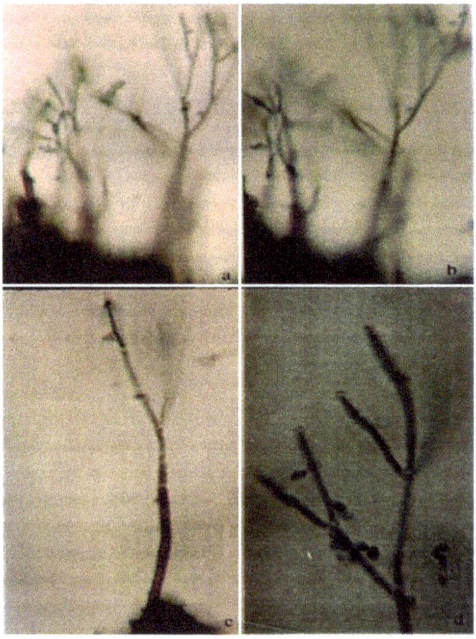

Fig. 45. *Ramichloridium musae*
a-d, Conidiophores with conidiogenous cells and conidia (450 x, 1000 x).

Colonies effuse, hairy, cottony, furinose to lanose. Mycelium submerged, septate, branched, hyaline, smooth, thin walled 1-2μm wide. Conidiophores macronematous, mononematous, older conidiophore dark, thick walled whereas proliferating younger conidiophores thick walled, erect, branched 35-150μm long. Conidiogenous cells terminal, polyblastic, straight, cylindrical golden brown, variable in length, cicatrised sympodial, 15-30μm long, 1.5-3μm broad. Conidia hyaline to sub-hyaline, thin walled, smooth, ellipsoidal, 3-5μm long, 2-3μm broad with inconspicuous basal scars.

On decaying leaves as well as rotting twigs from Papikonda, Bhimavaram, Bhadrachalam, Ashwaraopet forests of Khammam district, A.P., HCIO No: 43,015. NCBL No. 216, 319. Nov 2001. Coll. V.R.& G.S.K.

41. SACCARDAEA Cavara

Atti. ist. Bot. Pavia, 3: 346-347, 1894

Cavara in 1894, proposed *Saccardaea* with *S.echinocephala* as its type species. It produces synnematous, erect, pin head like colonies. The synnema are distinctly stiped with rounded or capitate heads containing setae and conidiophores. Mycelium submerged, septate, branched incorporated in the formation of synnemata. Conidiophores macronematous, synnematous, hyphal threads adpressed to form a long stipe. The heads of the synnemata spherical round or globose containing, conidiogenous cells and setae thread like or so, conidiogenous cells monophialidic. Conidia one celled, globose or elliptical.

Saccardaea was originally collected from Italy but this fungus was found to grow as a saprophyte on grasses and other monocots in Argentina, Austria, England and Italy. Mason and Ellis (1953), while studying *Periconia* they have examined other taxa such as *Stilbum, Sporocybe* and *Graphium* materials. On the basis of conidial development they have excluded these taxa and placed them as *Saccardaea atra* (Desm.) Mason & Ellis comb. nov. In the world there are three species of *Saccardaea.* Recently Tulloch (1972), monographed the genus and suggested that *Saccardaea* Cavara as a nomen dubium. He transferred *S.atra* to *Myrothecium,* but Carmichael et al. (1980) did not accept it as nomen dubium and listed them in the genera of hyphomycetes. Similarly, Hawksworth et al. (1995), mentioned it as nomen dubium. The authors while studying the fungi from Khammam came across this genus following Morris (1963), is being described here tentatively.

Saccardaea echinocephala, Cavara in Atti. ist. Bot. Pavia. 3: 346, 1894. (Fig. 46)

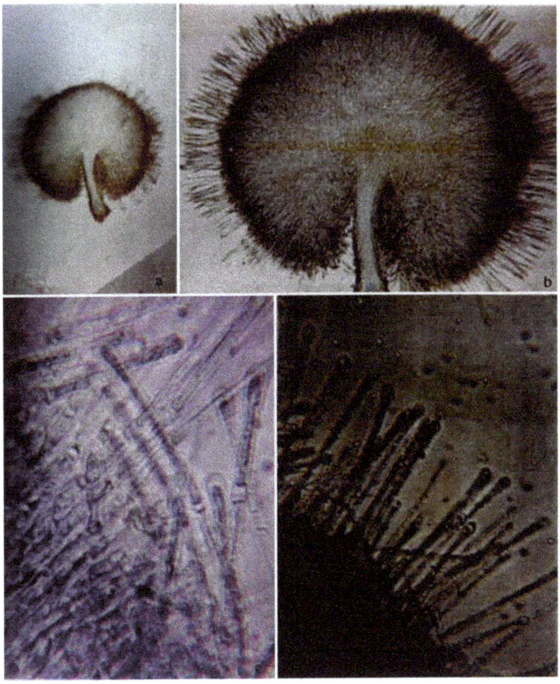

Fig. 46. *Saccardaea echinocephala*
a-d, Conidiophores, conidiogenous cells and conidia (100 x, 450 x, 1000 x).

Colonies synnematous whitish or pinkish pin head like mostly solitary or in tufts. Mycelium submerged in substratum, septate, branched, 2-3μm broad, hyaline to sub-hyaline, synnemata divisible into base, stipe, the surface of the stipe or named by variable shaped cells and capitate radiating head based part made up of thick walled and thin walled hyphae, stipe made up of closely, adpressed hyphae. Conidiophores which are macronematous, synnematous, simple in the stipe but branched in the capitate head, 2-3μm broad, thin walled, smooth but the textura of the synnema intricate made up of swollen cells. The conidiogenous cells phialidic, arising at the tips of conidiophore, very closely adpressed nearly, inseparable setae over growing conidiogenous cells radiating, thin walled, slightly swollen at the apex viewed under oil immersion or high power. Conidia enteroblastic arising from the phialides cylindrical, straight, 2-3μm broad, 10-15μm long, which become branched and distinct like hymenial layer of Agaricales each conidium globose or slightly elliptical 1-2μm in diam. No conidial germination and secession was observed.

Material from leaf litter from water bodies in Bhadrachalam, Kothagudem forests of Khammam district, A.P., HCIO No:43,014. NCBL No. 259, 313. Nov./Dec. 2001. Coll. V.R.& G.S.K.

42. SCLERODERMA Persoon

Syn. Meth. Fung. 706, 1801

Scleroderma was introduced by Persoon in 1801, with *S. citrinum* Persoon which was also known as *S. aurantium*. It is recognised by roughly oval basidiocarp but uneven hard and sessile or sub-sessile fruit body arising from cord like mycelium. Its outer membrane is peridium, which is thick olive yellow to brownish, scaly often in a net like manner, gleba greenish purple black spore mass with black purple tinge eventually escaping through the crack in peridium blackish brown, verrucose or reticulate, globose.

Seleroderma citrinum Persoon, is usually known as earthball. It grows commonly on the ground most often on bare soils in woods and wastelands. There are 25 species in the world. 18 species are known from India according to Bilgrami et al. (1991), and is distributed in various forests of North and South India.

Scleroderma citrinum Persoon ex Soothill and Fairhurst, The New Field Guide to Fungi Pp. 168-1978. (Fig. 47)

Fig. 47. *Scleroderma citrinum* a-h, Basidiocarps and gleba with basidiospores, i-k, V.S. section of basidiocarp with basidiospores (450 x, 1000 x)

Basidiocarps about 5-10 *cm* in diam., growing on the ground most often base soils in the woods and waste lands. The outer wall peridium which is slightly coloured the fruit body arises from cord like mycelium coming from the soil. The mycelium smooth, septate, with clamp connections or sometimes forming mycorrhiza with surrounding trees, grasses and plants. Fruit bodies or basidiocarp appear in August-December or earlier January later on get destroyed naturally by temperature, water and wind. The veritcal section shows outer thick peridium madeup of mycelium and skeletal cells central mass known as gleba with a system of mass of sterile veins there are no columella, capitilium, the basidiospores arise on the basidia which are seissile and autodigested and the basidiospores are freely suspended globose or spherical, 10-15μm in diam., verrucose or reticulate pinkish or blackish brown with purple tinge. Soothill and Fairhurst (1978) record that the basidiocarps are visible from August - December whereas they have been observed only in the month of November on wards. The peridium ruptures irregularly to release the spores.

From Bhadrachalam, Adilabad & R.R. district of A.P., NCBL No. 179, 212. Nov 2001. Coll. V.R.& G.S.K.

44. SCOLECOBASIDIUM Abbott

Mycologia 19: 29-31, 1927

Abbott (1927), introduced *Scolecobasidium* with *S. terreum* Abbott, as its type species, and further added another species *S. constictum* Abbott. *Scolecobasidium* produces effuse grey to brown flat or funiculose colonies. Mycelium superficial or immersed septate, branched, sometimes producing chlamydospores, straight or flexuous 2-3 celled, cylindrilcal. Conidiophores macronematous. Conidiogenous cells polyblastic, integrated, terminal or intercalary, sympodic, cylindrical or clavate or cuneiform, denticulate, denticles prominent. Conidia holoblastic, fusiform, ellipsoidal, cylindrical rounded at ends, sometimes T or Y shaped smooth or echinulate, 0-3 septate, pale or mid brown.

Scolecobasidium was isolated from soil, air decaying leaves latex, *Borassus* and lentils, roots of *Hevea, Vitis* from Congo, Egypt, India, N. America, Malaya, Srilanka, Australia, Japan and other places. It is mainly a saprophytic conidial fungus. Ellis (1971), has given an account of five species, described from various parts of the world. Matsushima (1971 & 1975), recorded *S. constrictum* Abbott, *S. humicola*, Barron et Busch, *S. terreum* Abbott, *S. variabile* Barron and Busch, from Japan and Papua New Guinea Islands. From India all the five species described by earlier workers were reported from U.P., Delhi, Karnataka, Kerala, Rajasthan and other places from soil and plant material (Bilgrami et al. 1991) in addition to those of Dwivedi (1959) *S. indicum*, Lodha (1970) *S. verruculosum* and *S. macrosporum* were added.

Scolecobasidium variabile Barron & Busch, Can.J.Bot.40: 83-84, 1962.

(Fig. 48)

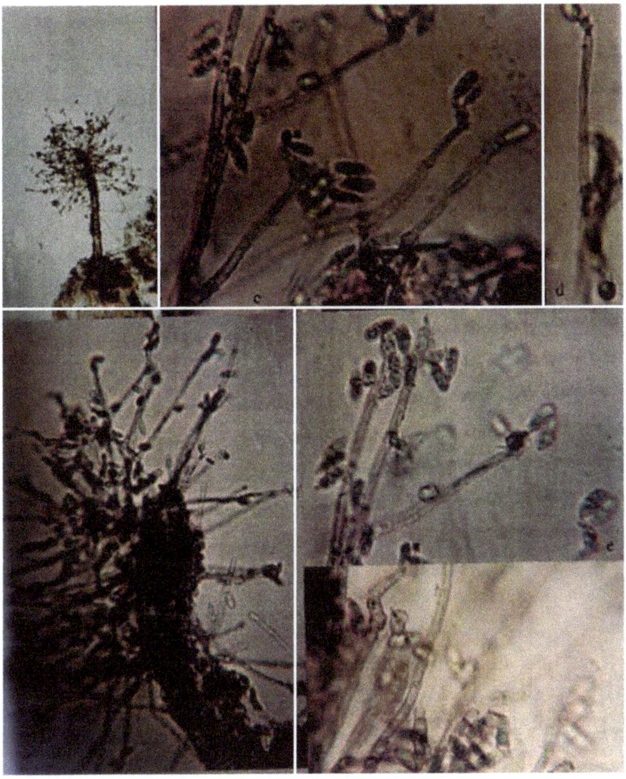

Fig. 48. *Scolecobasidium variable* - a, Colony. b-e, Conidiophores, Conidiogenous cells and conidia (100 x, 450 x, 1000 x)

Colonies effuse, slow growing, flat or raised or funiculose grey to brown, 1-2 *mm* across. Mycelium superficial or partly immersed, septate, hyaline, pale brown. Smooth or thick walled, 2.8-7.0μm broad. Chlamydospores formed in the natural substrate, thick walled irregular in shape 10-20μm diam. Conidiophore macronematous, mononematous, often short unbranched straight or flexuous, 2-3 cellled, cylindrical or variable in shape, 10-30μm long and 2.8-4.2μm broad. Conidiogenous cells polyblastic, integrated, terminal or intercalary sympodial, cylindrical, clavate or cuneiform, denticulate, 5-7μm wide and 10-20μm long, denticles cylindrical upto 1-4μm long, light brown coloured or uncoloured with the conidiogenous cells. Conidia solitary, acropleurogenuous ellipsoidal, oblong, cylindrical rounded at the ends, finely verrucose, constricted at the septate, 7-21μm long and 2.8-4.2μm broad. Conidial secession rhexolytic, conidial germination from the base.

From decaying unidentified leaves and bark from Bhadrachalam, Aswaraopet forests of Khammam, Vishakapatnam districts, A.P., HCIO No: 43,013. NCBL No. 133, 218. Oct. 2001. Coll. V.R.& G.S.K.

Scolecobasidium is widely known as foliicolous fungus but its isolation from soil and air is also common. *S. variabile* was first reported from N. America (Canada).

44. SPEGAZZINIA Saccardo

Michelia 2: 37, 172, 1880

Spegazzinia was proposed by Saccardo, with *S. ornata* Saccardo, which was similar to *Spegazzinia tessarthra* (Berk & Curt) Saccardo, is the present type species. *Spegazzinia* produces sporodochial colonies, black or brown in colour.Mycelium anastomosing, hyphae septate, branched, sub-hyaline to dark brown, smooth, thin or thick walled. Conidiophores micronematous, mononematous, basauxic. Conidiogenous cells globular, spherical, elongated, dehiscing in the median region and elongating intercalarily producing terminal, solitary, holoblastic. Conidia blastic, cruciform or 4-many celled, regular or irregular in shape, dimorphic spiny or smooth.

Spegazzinia is known in the world by six species and have been reported from U.K.Africa, Canada, India and Australia. Roquebert (1978), discussed the conidiogenesis at length which was later supported by Cole and Samson (1979). According to them *Spegazzinia* produces conidiogenous cells which are also known as conidiophore mother cells which produces one terminal conidia by holoblastic mode of conidiogenesis and later the outer wall of conidiophore mother cell breaks transversely and its inner wall elongates intercalarily to raise the height of the conidia. Each conidium carries a portion of conidiophore mother cell as freely as along with it. Subramanian (1956), described *Spegazzinia sundara* from India which produces 4 celled dimorphic conidia namely smooth and spiny conidia.

Spegazzinia intermedia Ellis, More Dema. Hypho. 475-476, 1976.
(Fig. 49)

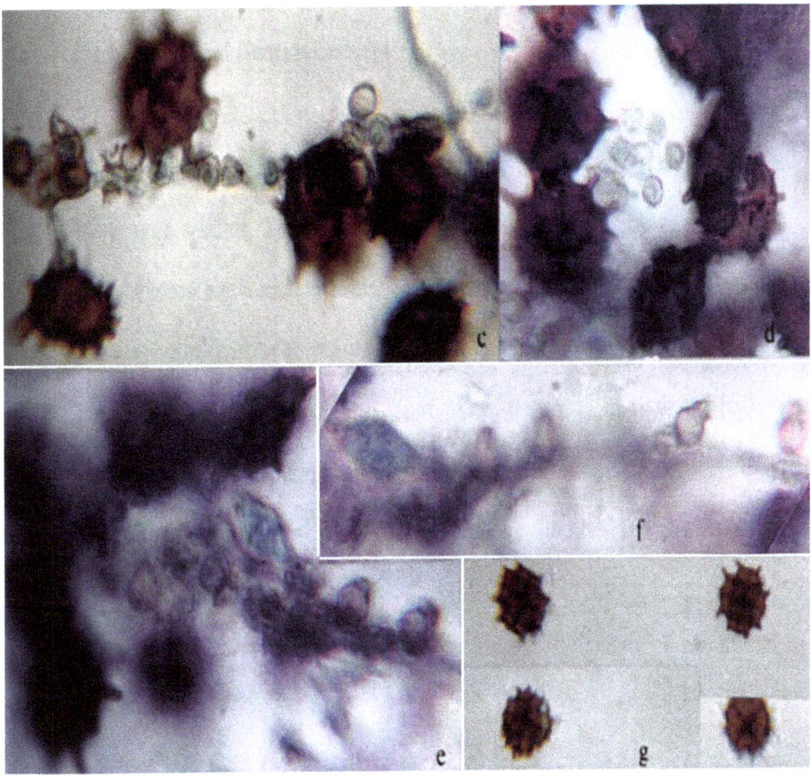

Fig.49. *Spegazzinia intermedia* - a-f, Colonies, conidiophores,, conidiogenous cells and conidia, g. conidia (450 x, 1000 x)

Colonies effuse, granulate, stromatic, spreading 2-3 *mm* across on the substrate. Mycelium anastomosing forming a network of variable cells in size and shape and thickness 3-7μm broad. Conidiophores (Conidiophore mother cells) micronematous or semi-macronematous bearing first holoblastic, 4 celled conidia, 5-8μm long, 3-5μm broad, upto 3-4μm broad at the base, 3-4μm broad at the apex, elongating intercalarily 15-20μm long, 2-3μm broad at base 3-4μm broad at apex, one celled. Conidiogenous cells terminal, holoblastic the conidiopohore mother cells tip acting like conidiogenous cells indiscrete terminal. Conidia holoblastic 4 celled, cruciform, each cell looking like a sector of sphere contricted at the septa and producing two spiny sharp projections regularly. The conidial dimentions 5-7μm in diam. No conidial germination and secession was observed probably it is schizolytic.

From leaf litter, decaying bark, twigs from Warangal, Adilabad, Bhadrachalam, Ashwaraopet forests of Khammam district, A.P., HCIO No : 43,032. NCBL No. 120, 311. Nov 2001. Coll. V.R.& G.S.K.

Spegazzinia intermedia Ellis, is new record to Indian mycoflora.

45. SPOROSCHISMA Berkeley & Broome

Gard. Chron. Pp. 540, 1847

Sporoschisma was proposed by Berkeley and Broome in 1847, with *S. mirabile* as its type species. *Sporoschisma* produces effuse stromatic colonies with phialophores, and capitate setae. Mycelium septate, branched, thick walled. Conidiophores (phialophores) macronematous, mononematous, erect in fascicles simple lageniform or cylindrical stout. Conidiogeonus cells monophialidic, integrated, terminal, percurrent, cylindrical, lageniform with long collarate neck. Conidia phialoconidia catenate, phragmosporous, cylindrical, septate.

Sporoschisma is well represented by its distribution from the work of Ellis (1971), Hughes (1966), Nagraj and Kendrick (1975), Matsushima (1975), Mercado et al.(1997). Bhat and Kendrick (1993) and Hughes(1996), attempted to redescribe the genus and accepted only 3 species namely *S. mirabile*, *S. nigroseptata* and rejected or transferred other species to *Charlara* etc. Holubova-Jechova and Hennebert (1972), described at another genus known as *Sporoschismopsis* which is characterised by proliferation of the conidiophore, and characteristic catenulate claviform, obovoid or pyriform conidia. *Sporoschisma* reminds ideal species described by Rao and Rao (1964), which is unique by its dark septate, erect, unbranched lageniform stalked phialophore and capitate setae which may be few in number. Recently, Bhat and Kendrick (1993), proposed *Sporoschisma uniseptata* from Andaman Islands.

Sporoschisma nigroseptata Rao and Rao, Mycopath. Mycol. Appl. 24: 81-84, 1964. (Fig. 50)

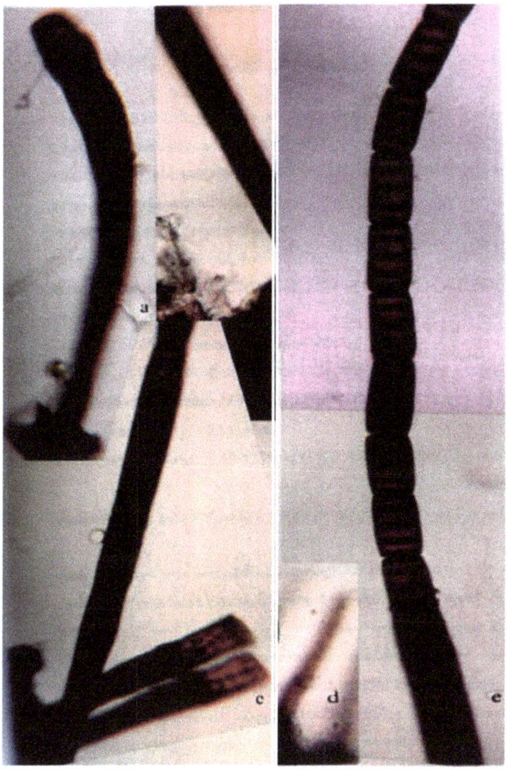

Fig. 50. *Sporoschisma nigroseptata* - a-e, Phialophores (conidiophores) capitate setae and conidial chain (450 x)

Colonies effuse, dispersed black, velvety, 1-2 *mm*. Mycellium immersed in the host tissue to form stromatic body. Stroma made up of thick walled cells raised above the substrate level producing phialophores immixied with capitate setae. Setae erect, unbranched, 5-6 septate, smooth 30-70μm long with capitate mucilaginous head. Conidiophores macronematous, mononematous, prominent, black, smooth with a swollen venter like part. Conidiogenous cells phialidic, enteroblastic with long neck and inflated venter producing conidia endogenously. The size of the phialide 200-270μm long, 15-20μm broad and 100-150μm long, collarate or neck. Conidia phialidic, 3-5 septate, 20-5μm long and 10-15μm broad septa dark walled the basal and apical hyaline in a chain later on getting seperated smooth, dark blackish brown. Conidial secession schizolytic germination could not be seen.

Leaf litter and decaying plant debris from Nizamabad, Adilabad, Medak, Bhardrachalam, Kotagudem forests of Khammam district, A.P., HCIO No : 43,033. NCBL No. 49, 169. Oct. 2001. Coll. V.R. & G.S.K.

46. STACHYBOTRYS Corda

Icones Fung. 1: 21f. 278 B., 1837

Stachybotrys was introduced by Corda in 1837, with *S.atra* Corda, as its type species. It produces effuse, black, colonies.Mycelium creeping septate, branched, producing erect, macronematous, mononematous, simple or branched. Conidiophores which were also known as phialophores earlier. Conidiogenous cells monophialidic, phialides hyaline or dark colour, variable in shape in clustered closely at the apex of the conidiophores, conidia phialidic one celled, variable in shape, solitary, sometimes clustered in mucilaginous mass.

Stachybotrys is known from all parts of the world by its 10 species. Bisby(1943), gave a detailed account of this genus and later Bisby(1945), considered *Memnoniella* and *Stachybotrys* as similar fungi because *Memnoniella* agrees in all respects the characters of *Stachybotrys* except that they produce conidial chains(*Memnoniella*), solitary or aggregated mass in *Stachybotrys*. Zuck (1946), demonstrated intermediate states between *Memnoniella* and *Stachybotrys*. Subramanian(1971),described *S.chartarum* (Ehrenberg) Hughes, *S.pulchra* Speg. and *S.nilagirica* Subramanian, as its species from various decaying plant material like *Musa* leaves, rhizophores of *Cajanus,* dead leaves of *Ficus* and twigs of *Altherosa.* Barron(1964),gave a modern account of this genus from Canada, Jong and Davis (1976), described a species of *Hyalostachybotrys* Srinivasan (pers comm.) which are nothing but dematiacious *Stachybotrys. Hyalostachybotrys* is a moniliaceous form therefore, it is synonym of *Stachybotrys* according to Barron (1964). Recently, Mckenzie (1991), studied *Stachybotrys* on *Freycinetia* sp., and reported number of species of *Stachybotrys* and proposed a key to the known and accepted species, Mckenzie (1991), took into account the species of *Stachybotrys* described till then. From India, Bligrami et al. (1991), recorded 18 species from various parts and substrates. However, recently Mercado et al. (1997), recognised and described 12 species of *Stachybotrys* known from Cuba and other parts of the world.

Probably, it is the latest excellent account of *Stachybotrys* available today. *Stachybotrys spherospora* from south Africa, Dorai and Vittal (1986), described *Stachybotrys ramosa* as a new species occurring on *Eucalyptus* litter from India.

1. *Stachybotrys nilagirica* Subramanian, Proc. Indian Acad. Sci.46: 331, 1957. (Fig. 51)

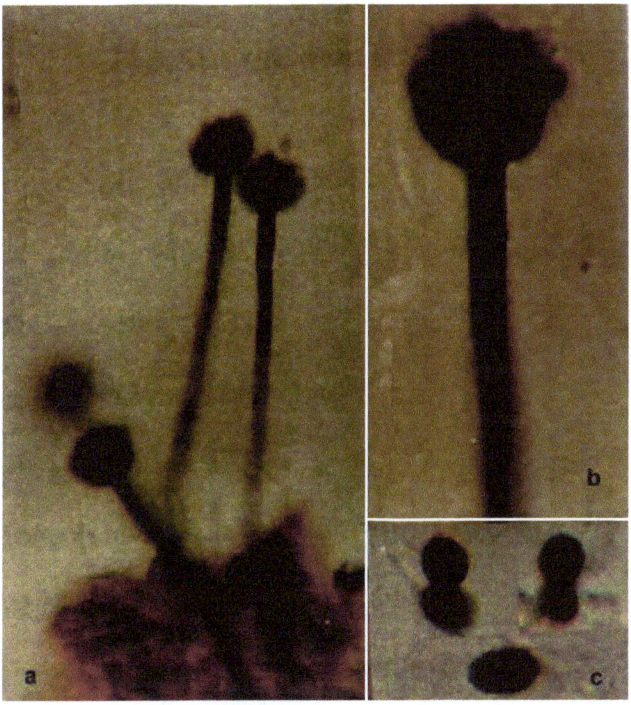

Fig. 51. *Stachybotrys nilagirica* - a-b, Colony and conidiophores conidiogenous cells, c. Conidia (450 x, 1000 x).

Colonies dispersed black pin head like. Mycelium superficial or immersed in the substratum to form small stroma, septate, thick walled, branched. Conidiophores micronematous, mononematous, erect, straight or flexuous. hyaline, sub-hyaline, septate, 100-300μm long, 5-15μm broad at base and 5-7μm broad at apex. Conidiogenous cells confined to the apex 5-7 in number, clustered ampuliformed, ellipsoid, colourless 10-20μm long, 5-10μm broad. Conidia enteroblastic, phialidic, one celled, globose, marked tuberculate dark greenish black. When mature 10-25μm in diam. No germination and secession of conidia was observed.

Leaf litter and decaying plants material like, bark, twigs etc., from Medak, Nizamabad, Bhadrachalam, Ashwaraopet of A.P. NCBL No. 29, 131. Oct. 2001. Coll. V.R. & G.S.K.

*2. **Stachybotrys pulchra*** Spegazzini, Rev.Age Y.Veter La Plata. Pp. 248, 1896. (Fig. 52)

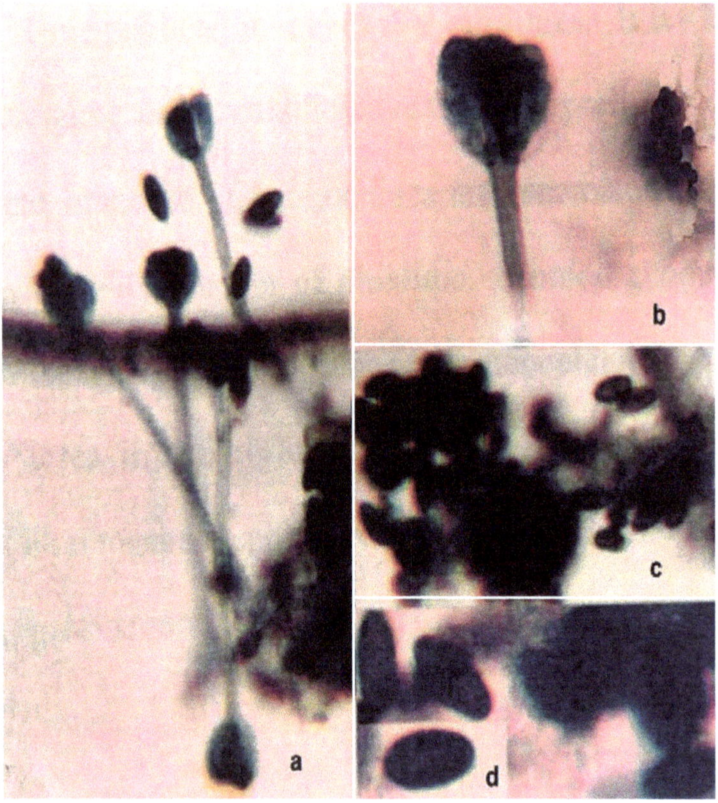

Fig. 52. *Stachybotrys pulchra*
a-b, Conidiophores, conidiogenous cells and conidia. c. Conidia (450 x, 1000 x)

Colonies gregarious, isolated, pin head like, irregular, yellowish black. Mycelium creeping septate, hyaline, smooth, branched. Conidiophores erect, macronematous, mononematous, straight, few septate, cylindrical, smooth, slightly swollen at base and rarely inflated at the apex, 100-200μm long, 5-10μm broad. Conidiogenous cells phialidic, ampulliform or inverted, short, bottled shaped, broad in the upper region and collarate at the apex, sometimes ellipsoidal, 5-7μm long, 3-4μm broad in the broadest below the apex. Conidia elongated or ellipsoidal, enteroblastic, 10-15μm long, 4-6μm broad, minutely granular one celled, dark black coloured.

Leaf litter and decaying plant material like, bark, twigs etc., from Bhadrachalam, Ashwaraopet, Yellandu of A.P., HCIO No:43,034. NCBL No. 151, 184. Oct. 2001. Coll. V.R. & G.S.K.

Stachybotrys is primarily a cellulosic fungus usually found on wet degrading paper materials, card board etc., and wet decaying plant debris.

Key to the species of *Stachybotrys*

I. Conidiophores macronematous, mono or synnematous.Conidiogenous cells phialidic at the tip. Conidia 1-celled some characters with conidial chains............................*Stachybotrys*

1. Conidia black, globose, verrucose................................*S.nilagirica*
2. Conidia black, elongated..*S.pulchra*

47. STACHYLIDIUM Link

ex Gray Mag. Ges. Nat. Freurde Berlin 3: 15, 1821

Stachylidium was proposed by Link 1809, with *S.bicolor* as type species. But for hyphomycetes starting point for vaild publication was 1821 therefore, its inclusion in S.F.Gray publication makes it a valid taxa. It is characterised by effuse hairy entangling mass, conidiophores. Its mycelium is superficial or immersed, septate, branched. Conidiophores macronematous, mononematous, simple or branched with number of phialidies. Conidiogenous cells arisining inverticils below the septa or conidiogenous cells monophialidic, discrete, arranged on the conidiophores, irregularly or inverticils cylindrical or ellipsoidal with collarate. Conidia acrogenous, phialidic, oval or cylindrical usually aggregated in mucilaginous mass around the mouth of the phialide.

Hughes (1958), while studying European herbaria in the light of experimental classification observed that Persoon's specimen was similar to the type, therefore, the proposed *S. verticillatum* (Persoon) Hughes, as its lectotype. *Stachylidium* is widely distributed fungus. It was reported from Commonwealth countries by Ellis (1971), and from Solomon Islands by Matsushima (1971), Barron reported it from the soils. Doi (1978), found out its ascogenous teleomorph from Japan. Recently, Mercado et al. (1997), described the *Stachylidium* from Cuba and recognised 3 species namely *S.bicolor, S.*var *caespitosa* and *S.cubense* from Cuba and many of its provinces.From India according to Bilgrami et al.(1991), *Stachylidium* was reported from various substrates like tea garden soils, stems of *Crotalaria* sp., *Tephrotia*, Tea flowers, *Phaseolus, Pyrus, Cannabis*, from Himachal Pradesh.

Stachylidium bicolor Link ex S.F.Gray, Link, Mag. Ges.Naturf.Fr.Berlin. 3: (1)15, 1809. (Fig. 53)

Fig. 53. *Stachylidium bicolor*
a-c, conidiophores with conidiogenous cells and conidia (450 x).

Colonies effuse, disperse, olive green, cottony, entangling mass of mycelium and conidiophores on the substratum. Mycelium immersed or superficial, septate, branched, sub-hyaline to pale brown, 2-3 *mm* broad, branched and forming a network,rarely stromatic bodies are seen. Conidiophores erect, macronematous, mononematous, branched, 200-500μm long, 5-7μm broad, septate, light brown to olivaceous brown, smooth at the base finally verrucose at the apex. Conidiogenous cells phialidic arising below the septa in verticils, cylindrical or ampulliform with collarette 7-15μm long, and 2-4μm broad, light brown, smooth, finely verrucose. Conidia enteroblastic, ellipsoidal to cylindrical hyaline to sub-hyaline occurring freely or aggregated in mucilaginous mass 3-5μm long, 1.5-3μm broad, smooth. No conidial secession or germination was observed.

On decaying leaf litter and barks, twigs from Nizamabad, Sri Kakulam, Bhadrachalam, Ashwaraopet, Yellandu forests of Khammam district, A.P., HCIO No:43,036. NCBL No. 240, 331. Nov. 2001. Coll. V.R. & G.S.K.

48. TETRAPLOA Berkeley & Broome

Ann. Mag. Nat. Hist. Ser. 5: 459, 1850

Tetraploa was created by Berkeley and Broome (1850), with *Tetraploa aristata* as its type species. It takes in forms producing effuse colonies blackish brown, hairy, mycelium superficial, septate and conidiophores micronematous, mononematous, conidiogenous cells intercalary indiscrete, monoblastic. Conidia muriform with 4 appendages which are setiform and tapering.

Tetraploa aristata and T. ellisii Cooke ex Ellis, were studied by Ellis (1949),in culture. Presently the genus is known from four more species namely *T.divergence* Tray and Earle, *T. mucicola* Cong ex Frag, *T. scabra* Harken. and *T. biformis* K. *Mastsushima* and T. Matsushima. *Tetraploa* is isolated from leaf litter in all temperature and tropical countries.Sree Ramulu and Ramalingam (1962), reported *Tetraploa* spores from air. Vasant Rao and Satyanarayana (1980), reported a strange character in *T. aristata.*

Tetraploa aristata Berkeley and Broome, Ann. Mag.Nat.Hist.Ser. 25: 59, 1850. (Fig. 54)

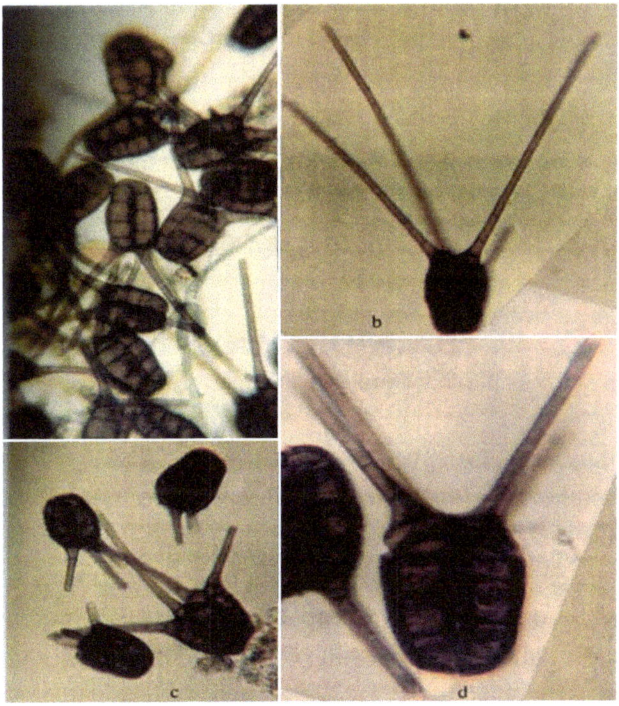

Fig. 54. *Tetraploa aristata* - a-d, Conidiophores and conidia (450 x, 1000 x)

Colonies effuse greyish brown black dots with projecting stiff hairs. Mycelium superficial pale brown septate 2-4μm wide smooth branch anastomosing. Conidiophores macronematous, micronematous short peg like. Conidiogenous cells integrated intercalary terminal or sometimes short lateral peg like structure. Conidia muriform or cruciform or urn shaped 4 rows cells. Each row consisting of 4 to 5 cells each. Each row forming a constriction in the configuration of the conidia, 10-14μm broad at base 10- 17μm broad at apex. 10-25μm long from each row one setae emerges apically. Thus each conidium may have atleast 4 setae but sometimes some of the conidia fail to produce the setae and grow to same length. They are variable inlength 2-3μm broad at base slowly tapering at the apex which is about 1-1.5μm broad, 3-10 septate conidia. In the muriform region black verrucose while in the setiform region it is smooth. Conidial secession rehxolytic germination from base and apex of the setae.

From leaf litter and plant debris from Bhadrachalam, Yellandu, Medak & Ranga Reddy districts, A.P., HCIO No:43,037. NCBL No. 141, 252. Oct. 2001. Coll. V.R. & G.S.K.

Tetraploa shows variable shape and variable lengths in the setae. In the specimens of Khammam two types of conidia are produced. Some conidia produce very short 2-3 setae where as others produce 4 setae of which one is small or the conidia get divided to 2 vertical half each carrying 2 setae and showing a clear division. Matsushima and Matsushima 1996, described a new species namely *Tetraploa biformis* in which 2 setae are short and curved where as other 2 setae are projecting upwards.

49. THAROOPAMA Subramanian

J. Indian Bot. Sere. 35: 83-86, 1956

Subramanian (1956), proposed *Tharoopama* with *T.trina* Subramanian, as its type species. *Tharoopam*a produces synnematous, erect colonies. Mycelium superficial or immersed in the substratum, septate, branched. Conidiophores macronematous, synnematous forming a conidiomata composed of more or less compacted group of erect broadly fused conidiophores parallel and splashing in the head region. Conidiogenous cells polyblastic, sympodial, geniculate, denticulate. Conidia hyaline, globose, one celled.

Tharoopama was reillustrated by Morris (1963), in his illustrated genera of synnematous fungi. Dev Rao in (personal communications) mentioned that *Tharoopama* was reported from A.P., as a new species, Reddy et al. (1975) reported *T.naimnagarensis* from Warangal on dead leaves of *Phoenix sylvestris*.

Tharoopama trina Subramanian, J.Indian. Bot.Soc.35:83-86, 1956. (Fig. 55)

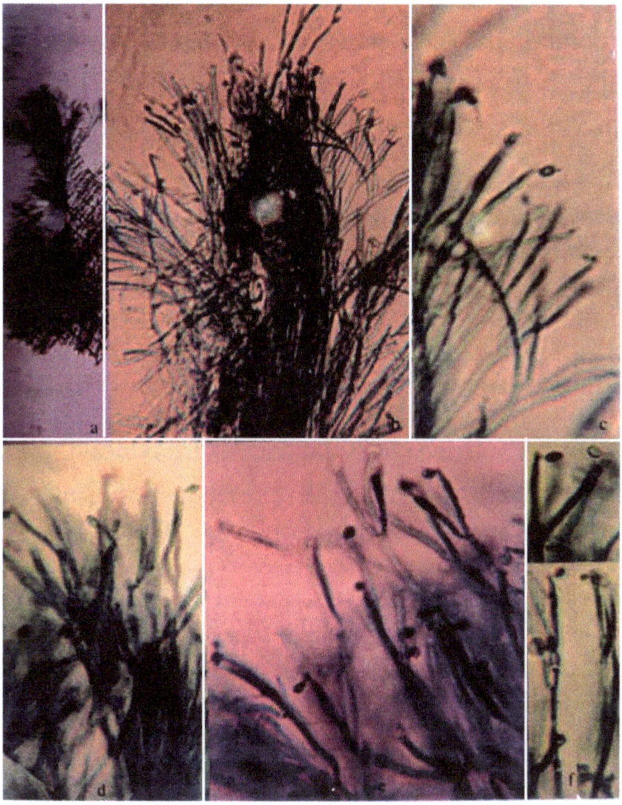

Fig. 55. *Tharoopama trina*
a-f, Colonies, conidiophores, conidiogenous cells and conidia (100 x, 450 x, 1000 x)

Colonies synnematous, scattered, superficial, erect, brownish black stalk and an apical radiating ash grey globose head or somewhat irregular in out line. Mycelium superficial or immersed in the host tissue, septate, sub-hyaline to pale brown, smooth, 2-3μm broad. Conidiophores macronematous, synnematous, 700-1500μm long, 50-100μm broad in the basal region, 500-1000μm broad in the head region, individual thread of the conidiophores aggregated paralleled textura conidiomata parallel smooth, septate, synnemata, progressively diverging from the main stalk to form a divergent to the head of the synnemata. Conidiophores which are the free ends of the hyphae of the synnema brown in colour but hyaline towards the tips diverging at various angles. Conidiophores branched terminally to give rise to the conidiogenous cell terminally. Conidiogenous cells terminal, polyblastic, geniculate, branching. The conidiogenous cells are cylindrical one celled 15-30μm long, 3-4μm broad, geniculate but the geniculations may not be seen in young synnema in which the conidial production has just started but are conspicuous in older ones. Conidia holoblastic, polyblastic with a basal apiculous hyaline to sub-hyaline one celled 3-4μm broad, 3-5μm long smooth. Conidial secession germination could not be observed. Conidiogenon cells similar to *Hansfordia* or *Nodulisporium* but differing from them in the conidial formation as discussed by de Hoog (1973). *Tharoopama* produces *Tritirachium, Acrodontium* like conidiogenous cells. *Hansfordia, Nodulisporium*, produce flat conidial base (Scar) where as *Tharoopama, Acrodontium, Tritirachium* they produced zig-zag sympodic cicatrised conidiogenous cell.

From leaf litter and decaying plant debris from Bhadrachalam, Yellandu forests of Khammam, Adilabad, Nizamabad districts of A.P., HCIO No: 43,038. NCBL No. 130, 200. Nov. 2001. Coll. V.R. & G.S.K.

Tharoopama is known to A.P., and India by its 3 species but the earlier collections were made from Adilabad, Nizamabad, Warangal and Khammam.

50. TOLYPOCLADIUM Gams

Persoonia 6(2): 185-191, 1971

Tolypocladium was created by Gams, with *T. inflatum* as its type species. He has also discussed other species namely, *T. geodes*. Gams, and *T. cylindrosporum* Gams, in the same paper. *Tolypocladium* produces diffuse colonies which are white to ochraceous. Mycelium cottony, thin walled. Condiophores mononematous or micronematous. Conidiogenous cells phialidic with inflated base and long tapering neck producing phialoconidia which are globose to cylindrical, hyaline.

Tolypocladium was reported earlier from Netherlands but it was isolated from pinus species in Canada. Holubova-Jechova (1988), described about 10 species and proposed a key for identification. Riba et al. (1986), tried to study its isozyme analysis. von Arx (1986), and Rakotonirainy (1991), suggested its comparison with *Beauveria* in rRNA sequence. Stimberg et al. (1992), studied the karyotyping with electrophoretic methods. Thus suspect a alkaloids cyclosporin in it. Initially this fungus was mistaken for *Trichoderma*. A serious review of the literature revealed that it is not *Trichoderma* because of its sparse conidiophores and little secretion of mucilage to hold the conidia in the mass.

Tolypocladium geodes Gams, Persoonia 6(2): 187-188, 1971. (Fig. 56)

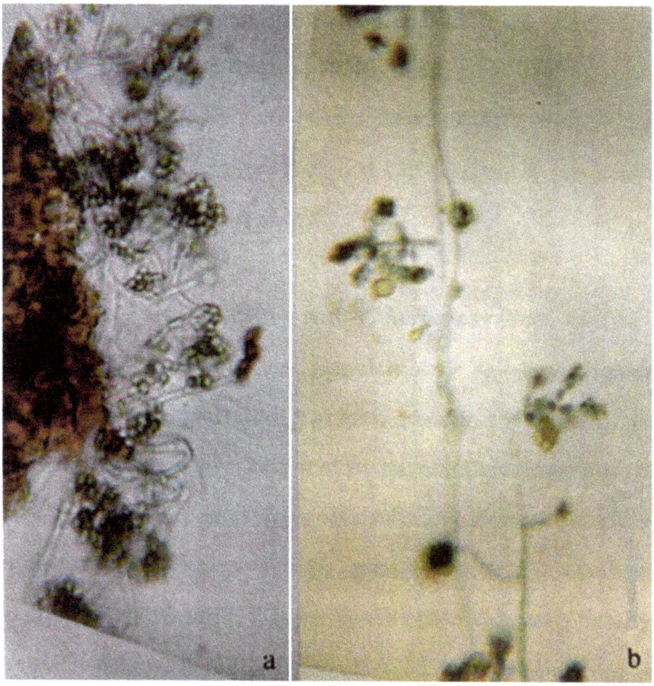

Fig. 56. *Tolypocladium geodes* - a-b, Colony, conidiophores, conidiogenous cells and conidia (450 x).

Colonies effuse, ochraceous, whitish. Mycelium superficial creeping cottony, thin walled, 2-3μm broad, branched, delicate. Conidiophores macronematous, mononematous, simple or branched lateral on the mycelium

10-30μm long, bearing terminally, laterally or inverticils of the conidiogenous cells. Conidiogenous cells phialidic which may be straight or curved, thin walled, 10-20μm long, 2-3μm broad at base and 1-2μm broad at apex, cylindrical or elongated, long, bottle shaped with straight or bend neck. Conidia phialidic, spherical, 1-3μm in diam., held in loose mucilaginous mass temporarily and later getting separated. No conidial attachment, secession and germination was observed.

From leaf litter and plant debris from Bhadrachalam, Yellandu, forests of Khammam district, A.P., HCIO No:43,039. NCBL No. 161, 336. Nov. 2002. Coll. V.R. & G.S.K.

Tolypocladium is a very important fungus, a potential to be used in pharmaceutical industries in the production of antibiotics and alkaloids.

51. TORULA Persoon

ex Link, Mag. Ges. Nat. Fre. Berlin, 3: 21, 1809

Torula was proposed by Persoon (1801), with *T. herbarum* (Pers. ex Fries), as its type species. *Torula* produces effuse, granular, black powdery colonies on the substratum. Mycelium superficial, toruloid, septate, cylindrical, branched. Conidiphores micronematous, terminal or lateral. Conidiogenous cells are one-celled, discrete, terminal or intercalary, polyblastic, cup shaped. Conidia 0-many septate in chains when phragmo-conidia, there is a distinction between basal and apical cell, the apical cell if it is partly thickened and partly sub-hyaline it acts like a conidiogenous cell to another conidium to produce a simple or branched chain when the apical cell turns thick walled it become sterile and it does not produce any further conidium.

Torula Persoon, is the very old established fungus and there are number of species which are heterogenous and are placed under the same name. Many workers consider *Bahusandhika* species *B. caligans* Batista and others similar, but the author considers it as separate distinct form. Vasant Rao and de Hoog (1975), have studied number of cultures under the name *Torula* and properly segregated the cultures to their respective genera.

Torula herbarum Pers ex Link 1809- Mag. Ges. Naturf Freunde, Berlin. 3: 21. (Fig. 57)

Fig. 57. *Torula herbarum* - a-d, Colony, conidiophores, conidiogenous cells and conidia (450 x, 1000 x)

Colonies effuse about 3-5μm in diam., and about 1 *mm* high blackish brown. Mycelium superficial thin or thick walled sub-hyaline to reddish brown, 1.5-2μm broad, septate, branched. Conidiophores terminal or lateral integrated, indiscrete, micronematous, mononematous. Conidiogenous cells thin walled, fertile, where as lower part thick walled, sterile, 3-6μm tall and upto 5μm broad, may be terminal or the apical cell of the phragmo-conidium or lateral. Phragmo-conidia 3-4 celled pear shaped or conical, deeply constricted at the septa in catenulate chains holoblastic, apical cell fertile produces another conidium. If it is sterile it becomes dark thick walled, 30-40μm long, 10- 15μm broad at apex, sterile conidia 3-4 celled, terminal cell thick walled not producing conidia.

From decaying plant debris from all over Andhra Pradesh, HCIO No:43,040. NCBL No. 177, 222. Nov. 2002. Coll. V.R. & G.S.K.

Matsushima (1971&1976), reported number of species from Japan and observed that some of his species do not agree the generic concepts and therefore, are not accepted by later workers. Hashmi et al. (1973), studied the conidial ontogeny and they consider *Torula* as a blastoconidial fungus whereas Ellis (1971), observed it as treto conidial. The authors failed to observe any treto conidia and agree with Subramanian (1971), who considered it as blastosporic fungus.

52. TRETOPILEUS Dodge

Bull. Torr. Bot. Cl. 73: 223, 1946

Dodge (1946) introduced *Tretopileus* with *T. opuntiae* Dodge, as the type species. Deighton (1960) while studying the African fungi from Sierra Leone and materials of Congo collected and recorded by Berkeley & Curtis, examined this fungus.

Further Deighton with the help of Dr. S.J. Hughes and Mr. EW Mason exasmined the fungus of Berkeley & Curtis and proposed a new combination viz., *Tretopileus sphaerophorous* (Berk. & Curt) Hughes & Deighton which is conspecific with the typic species. Therfore, according to the Code (Staflau 1974) *T. sphaerophorus* (Berk.& Curt.) Hughes & Deighton becomes the type species. *Tretopileus* takes in forms which produces effuse, colonies, mycelium is superficial or immersed, hyphae septate, stroma present, conidiophores synnematous macronematous (gemmiferous stipe), conidiogenous cells (gemmigenous cells) indiscrete, indeterminate, terminal or intercalary, percurrent or annellated, conidia (gammae) with obscure origin compsite, spherical or disc-shape muriform, dark blackish brown. *Tretopileus* occurs on different host ranges like Congo beans, *Cajanus, Opuntia* and several unidentified substrates like barks. It was reported earlier from India, in Maharashtra by Patil (1964) and from Orissa by Reddy (1983), Carmichael et al. (1980) considered it under Mycelia sterilia.

Tretopileus sphaerophorus (Berk & Curt.)Hughes & Deighton, apud. Deighton, Mycol. Pao. 78 : 2-4, 1960. (Fig. 58)

Fig. 58. *Tretoprileas sphaeroplorus* a-b, Colony, gammiferous stype, with gammae (450 x)

Colonies effuse, tufted, black, visible to naked eyes, tall, rough, pin- head like. Mycelium immersed in the host tissue, hyaline, forming stroma. Stroma submerged or erumpent above the surface of the substrate made up of thin-walled cells (prosenchyma) giving rise to conidiophores. The conidiophores also as gemmiferous stipe, erect, straight or flexuous tough, made up of several hyphae threads, synnematous, 80-700μm long and 35-70μm thick bearing characteristic annellations or ring- like structures, sub-hyaline to pale brown, smooth, septate, unbranched rarely branched. Conidiogenous cells (gammigenous cells) obscure giving rise to gammae in unknown mode but in young condition hyaline protrusions arise all over the rim of the tip of conidiophore (stipe) the conidiophore grows even after giving of the conidia or gammae. The conidia solitary, terminal, acrogenous, origin not known, obconoid with a truncate or slightly concave, 60-120μm across and 30-60μm thick, dark blackish brown, pellucid spots present. Germination of the gamma could not be observed.

Collected on *Delonix elata* bark Khammam, Adilabad, Medak, Srisailam, Nalgonda, A.P., NCBL No.90, 180, 271,299, Oct. 2002. Coll. V.R. & G.S.K.

53. TRICHODERMA Persoon

Syn. Meth. Fung. 230, 1801

Persoon (1801), introduced *Trichoderma,* a cosmopolitan fungus with *T. viride* as type species. However, prior to his work of 1801 Persoon (1794), had published another fungus under the name, *Trichoderma* which is a nominuin dubium a Myxomycete member. This fungus occurs on decaying leaf litter, plant debris and it has been isolated from the soil, mushroom compost. It produces greenish tufty coalescent colonies with whitish tinge.Mycelium superficial septate branched.Conidiophores macronematous, erect, monomematous branched. Conidiogenous cells terminal, lateral, enteroblastic, phialidic. Conidia one celled dry or slimy, smooth or encrusted with wart like protuberances hyaline or green.

Trichoderma is very important anamorphic conidial fungus and also known by its teleomorphic forms like *Hypocrea, Podostroma* and *Thuemenella* belonging to hypocreales order of Ascomycotina. It is known by its 2 species distributed in all parts of the world. It seems that it is most studied fungus in taxonomy and industrial microbiology, because of its antibiotic properties. It is employed in biological control, Rifai (1969), Barak et al. (1985), Birstt (1984), Hughes et al. (1993), found it in all parts of the world like U.S.A., Canada, S. America, African Countries, India, Srilanka, Japan and other countries. In India alone there are 12 species known which are isolated from *Cannabinus, Cassia, Cotton, Aesculus, Corchorus, Eucalyptus, Pennisetum, Potato, Nicotiana, Sorgham seeds, Arachishypogea, pods, Kernels etc., Zea, cichoriu, Cajanus, Thuja, Cuprassus* and *Quarcus* like host of different substrates in various states. The species of *Trichoderma* have used in applied sciences for cladistic analysis of isoenzyme polymorphism Stasz et al. (1989), for mitochondrial and plasmid DNA studies Meyer (1989), and for DNA finger printing and differentiation of stains Meyer et al. (1991). It is interesting to note that Gullino (Editor) arranged an international workshop on *Trichoderma* and *Gliociadium* and published the deliberation in Petria (1991), Leong and Berks (Eds) considered the significance of *Trichoderma* in Molecular industrial mycology. The crude usage of *Trichoderma* cultures beyond doubt has controlled several seed borne, and soil borne pathogens.Its cultures have been sprayed on standing crops to control air brone diseases. *Trichoderma* has enormous biochemical potentials to be used in different fields. Recently a monograph has been published by Nagamani et al. (2002).

1. *Trichoderma viride* Pers ex. Fr.Syst. Mycol. 3: 215, 1829.

(Fig. 59)

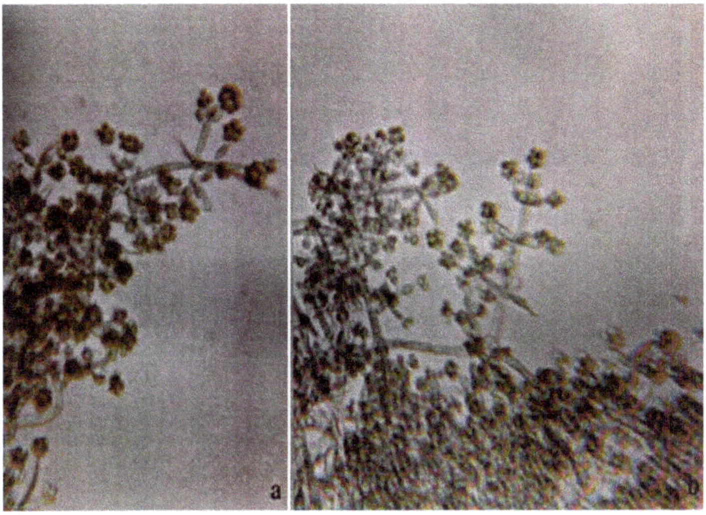

Fig. 59. *Trichoderma viride* - a-b, Colony, conidiophores, conidiogenous cells and conidia (450 x)

Colonies effuse, cottony, delicate whitish green in colour measuring 2-5 *mm* on the substrate very conspicuous, visible to naked eyes. Mycelium superficial, creeping, hyphae septate, branched, 2.8-5.6μm wide, anastamosing. Conidiophores, macronematous, semi-macronematous, mononematous, erect or branched 10-15μm long, 2.8-5.6μm broad. Conidiogenous cells enteroblastic, phialidic, one celled slightly narrow, broader base in middle and narrow at apex hyaline to sub-hyaline 5-20μm long, 2.8-4.2μm broad at base, 1.4μm broad at apex smooth, hyaline to sub-hyaline. Conidia zero-septate dry or slimy, solitaring or in chains or in cluster of conidial mass 2.8-4.2μm in diam., smooth or slightly verrucose bright green in colour. Conidial secession could not be observed.Conidial germination could not be seen.

On unidentified dead leaves and bark, plant debris from Bhadrachalam, Yellandu, Ashwaraopet forests of Khammam district, A.P., HCIO No:43,041. NCBL No. 58, 122. Nov. 2001. Coll. V.R. & G.S.K.

2. *Trichoderma koningii* Oudem in Oudemanus & Koning 1902 Arch. Neerl. Sci. Ser. 2, 7:291 (Fig. 60)

Fig. 60. *Trichoderma koningii*
a, Colony, conidiophores, conidiogenous cells and conidia (450 x).

Colonies effuse, greenish, spreading about 3 *cms* on natural substrate. Mycelium superficial, creeping on the substrate, septate, branched, 2-3μm broad, subhyaline to greenish brown profuse branching leading to the formation of the conidiophores. Conidiophores micronematous or semi-micronematous, one celled or septate, cylindrical, erect, simple unbranched the branching system of the mycelium gives an appearance of conidiophore branching the whole system looks like elongated pyramid. 3-5μm long, 1.5-3μm broad, smooth.Conidigenous cells phialidic, lagnilform or ampulifrom or spindle shaped, straight or curved with short or long neck irregularly borne directly on the hyphae or on the conidiophore with no order of development variable in shape and size. Conidia phialidic 0-septate, greenish brown, oval in shape, 2-3μm long upto 2μm broad, smooth. No germination was observed in the present study.

On decaying plant debris from Bhadrachalam, Yellandu, Aswaraopet forests of Khammam, Hyderabad, Warangal districts of A.P., HCIO No:43,042. NCBL No. 77, 101. Nov. 2001. Coll. V.R. & G.S.K.

3. *Trichoderma harzianum* Rifai, Mycol. Pap. 116: 1-56, 1969. (Fig. 61)

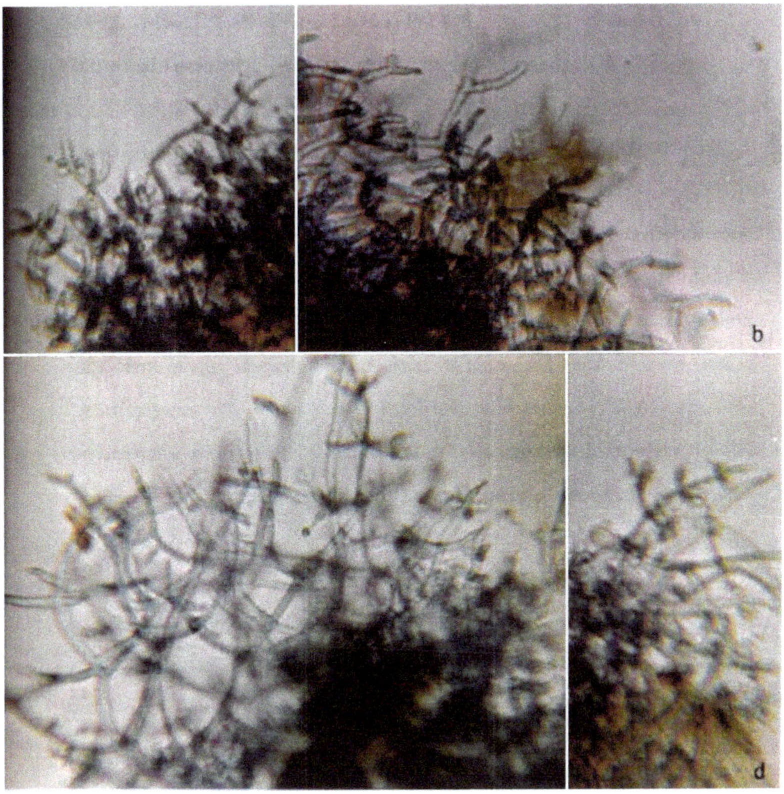

Fig. 61. *Trichoderma harzianum*
a-d, Colonies, conidiophores, conidiogenous cells and conidia (450 x).

Colonies effuse, greenish brown. Mycelium creeping, septate, branched, hyaline to sub-hyaline, smooth, 2-3μm broad, septa 10-15μm apart. Conidiophores micronematous or semi-micronematous, simple or branched, 20-30μm long, 2-3μm broad, producing phialidic. Conidiogenous cells phialidic terminal or inverticils on the conidiophores laginifroms with long neck, 8-12μm long, 2-4μm in the broadest upto 5μm broad in the pointed neck like region. Conidia greenish held in mucilaginous mass to from a head consisting 5-10 conidia.

From Bhadrachalam, Yellandu forests of Khammam, Hyderabad, Medak districts of A.P., HCIO No:43,043. NCBL No. 38, 166. Oct. 2001. Coll. V.R. & G.S.K.

T.viride is a very important fungus, in addition to *T. viride* the presence of *T. koningi,* and *T.harzianum* species suggests that Khammam is rich in diverse species of *Trichoderma. Trichoderma* in general produces antagonistic and antibiotic substances. The abundant occurence of this fungus is suggestive of new industrial applications particularly in agricultural industry. *Trichoderma* is employed in seed treatment to control seed borne diseases and pathogenic fungi present in the soil. *Trichodema* is a vital micro-organism its potentialities and applications are yet to be worked out. The work is in active progress throughout the world and including India in the fields of agriculture, pharmaceutical, medicinal and industrial applications are yet to come. It is observed that *Trichoderma viride* can produce conidia in chains or in clusters like balls or solitary, at the tip of the phialide. The situation depend, mostly upon the atmospheric conditions i.e. saturation and humidity. Therefore, it creates confusion with its allied from *Gliocladium.*

Key to the species of *Trichoderma*

I. Conidiophores macronematous may end sterile, variable shape. Conidiogenous cells phialidic. Conidia in aggregated mass or solitary, 1-celled .... *Trichoderma*

1. Phialo-conidia in globose mass,1-celled roughened .... *T. viride*
2. Phialo-conidia 1-celled ellipsoidal or oblong, smooth.. *T.koningii*
3. Phailo-conidia 1 -celled globose or sub-globose.. *T.harzianum*

54. TRIMMATOSTROMA Corda

Icones Fung., 1: 9, 1837

Trimmatostroma was proposed by Corda 1837, with *T.salicis* Corda, as its type species. *Trimmatostroma* takes in sporodochial fungi which may be sometimes amphigenous in the host tissue, effuse, black, velvety often covering quite large areas. Mycelium superficial or immersed taruloid thick walled. Conidiophores micronematous or semi-micronematous arising in fascicles, meristamatic, pale brown. Conidiogenous cells terminal, intercalary. Conidia variable in shapes, branched in chains, several septate, smooth or verrucose.

Trimmatostromia is wide spread cosmopolitan found in all tropical countries. Its wide spread Ellis (1976), has reported a clear stroma in certain species. The material had been collected in Great Britain, U.S.A., S.Africa and from India Manoharachary et al. (1977), reported a new species which produces branched chains of conidia. Sutton (1973), Studied *T. betulinum* (Corda) Hughes, *T. salicis* Corda, from Canada on a number of host materials. In the taxonomy of *Trimmatostroma* there is little bit confusion with *Coniothicium betulinum* Corda *Fumago* and other taxon. *Trimmatostroma* has also been isolated from Indian soils is *T. terricola* Manoharachary et al. (1977), Ananthagiri, Visakhapatnam, A.P., India. In the studies on fungi from Khammam a species of *Trimmatostroma* collected which is being described here.

Trimmatostroma indica Manoharachary & Rama Rao, Raghuveer Rao, Curt, Scl. 46: 788, 1977. (Fig. 62)

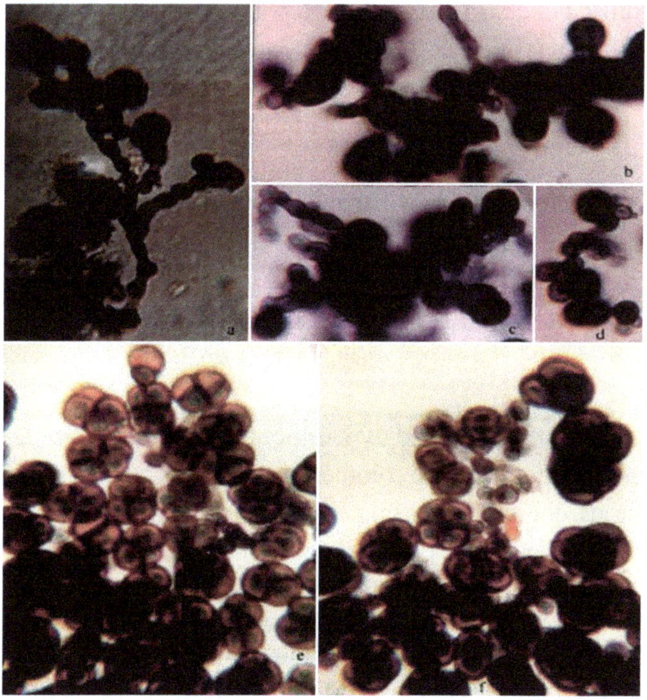

Fig. 62. *Trimmatostroma indica*

a-f, Conidiophores, conidiogenous cells and conidia (450 x, 1000 x)

Colonies sporodochial pulvinate, crumpet, blackish brown to black. Mycelium immersed, septate, branched, toruloid, 2-7μm broad, thick walled, sub-hyaline to pale brown. Stroma thick walled, prosenchymatous producing conidiophores. Conidiophores micronematous, or semi-micronematous, simple or branched, septate, sub-hyaline to dark brown, 15-17μm long, 3-5μm wide, septate, toruloid, constructed at the septal thick walled. Conidiogenous cells terminal of lateral monoblastic or polybiastic or intercalary discrete or indiscrete when discrete ampulliform arising on the creeping hyphae. Conidia holoblastic 2-4 celled or more transversely septate, thick walled, dark brown to black, smooth or verrucose produced in simple or branched or irregular chains 25-45μm long, 10-20μm broad. Conidial secession schizolytic getting separated into conidial shapes of *Dictyoarthrinium, Circinella, Torula* etc.

Material from leaf litter from Bhadrachalam, Ashwaraopet, Nalgonda, Nizamabad districts of A.P., HCIO No. : 43,044. NCBL No. 176, 289, Nov. 2002. Coll. V.R. & G.S.K.

Trimmatostroma presently described from Khammam forest is slightly different from *T. indica.* In this specimen the toruloid nature of the conidiophore and production of secondary conidia according to Subramanian in personal communication to Vasant Rao in 1993. The first conidium is holoblastic later conidium in which there is a short cylindrical cell which produces another conidium like *Dictyoarthrinium, Spegazzinia* etc., tentatively it is placed in *Trimmatostroma indica* Manoharachary only.

55. TRITIRACHIUM Limber

Mycologia 32: 16, 1940

Tritirachium was proposed by Limber (1940), with *T. dependens* as its type species.It produces effuse colonies of septate mycelium, macronematous, mononematous, conidiophores branched, branching inverticil. Conidiogenous cells geniculate, cicatrised, polyblastic. Conidia one subglobose to elliptical thin walled.

Tritirachium Limber was first isolated from soil in America, Europe, Canada, and Japan. It is superficially similar to *Verticillium.* Therefore, de Hoog (1972), studied such genera like *Verticillium.,*

Tritirachium, Beauveria, Geniculospora, Nodulosporium and proposed *Acrodontium,* de Hoog as new genus *T. roseum* has been reported from India by Rama Rao and Manoharachary (1990), from soil. Bilgrami et al. (1991), has reported at 5 species of *Tritirachium* from India it has been isolated from soil, dead stems of rice seeds of *Brassica* and in Goa, it has been isolated from human being causing skin disease. *Tritirachium* is to be identified very critically from other genera by its conidiogenous cells. The conidiogenous cells are geniculate wavy in *Tritirachium* similar to conidiogenous cell with denticles becomes *Acrodontium* or it condiogeonus cell produces conidia flatly then it becomes *Nodulosporium, Hansfordia* etc.

Tritirachium dependens Limber, Mycologia 32: 16, 1940. (Fig. 63)

Fig. 63. *Tritirachium dependens*
a, Colony, b-c, conidiophores, conidiogenous cells and conidia (100 x, 450 x).

Colonies effuse, hairy, branched, grey to whitish brown. Mycelium submerged in the host tisstie, thin walled, hyaline to sub-hyaline, 2-3μm broad. Conidiophores in a macronematous, mononematous, erect, 500-900μm or more long, 2-5μm broad at base and 2-3μm broad at apex, smooth, thin walled, septate sub-hyaline to pale brown. Conidiogenous cells polyblastic, terminal or lateral inverticils broader at base getting very narrower towards apex. 5-7 conidiogenous cells at each point produced on primary or secondary branches, 2-3μm broad and 50-100μm long. Conidia hyaline, slightly whitish to yellow, globose to elipsoidal. 2-3μm long, 1.5-2.5μm broad, smooth thin walled.

From decaying plant debris from Papikonda, Bhadrachalam, Yellandu, Ashwaraopet forests of Khaminam district, A.P., HCIO No:43,045. NCBL No. 226, 301. Dec. 2002. Coll. V.R. & G.S.K.

56. VERTICILLIUM Nees ex Wallroth

Fl. Crypt. German, 2: 301, 1833

Verticillium was proposed by Nees, with *V. tenerum* as its type species. It is one of the old established fungi known as a parasite or saprophyte isolated from the soil. Its colonies are effuse hyaline or lightly coloured. Mycelium creeping septate, branched, smooth thin walled. Conidiophores macronematous, mononematous, simple or branched, branching in various orders and inverticils. Conidiogenous cells phialidic on the tip of the ultimate branches, thin, cylindrical, narrowly tapering at the apex. Conidia phialidic and aggregated into mucilaginous mass.

Verticillium theobromae (Durc) Mason. Hughes, Mycol. Pap. 45: 10, 1951. (Fig. 64)

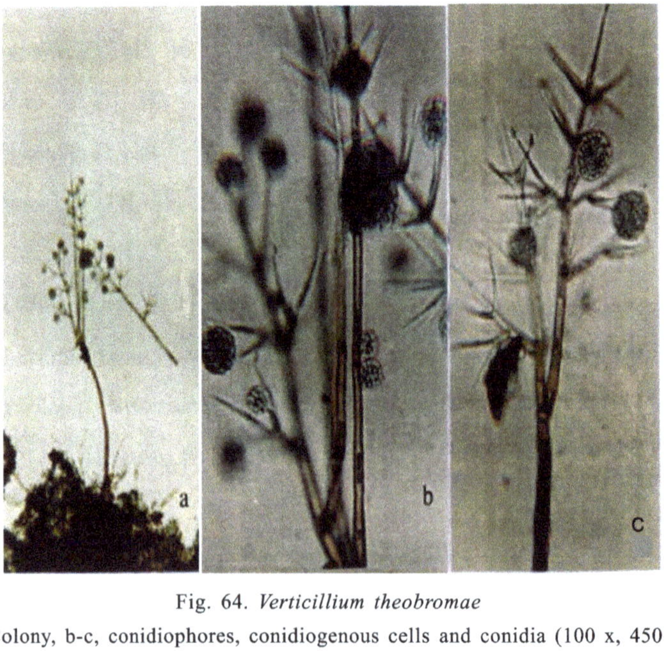

Fig. 64. *Verticillium theobromae*
a, Colony, b-c, conidiophores, conidiogenous cells and conidia (100 x, 450 x).

Colonies effuse, dispersed, hairy with small tufts of branches inverticils whitish or coloured. Mycelium immersed in the substrate, branched, septate, 2-3μm broad or sometimes broader in culture. It produces chlamydospores and stroma in nature, stroma thick walled, black in colour giving rise to conidiophore in solitary or in branches. Conidiophores macronematous, mononematous, simple or branched, septate, cylindrical, basal thick walled, dark black, 3-5 μm broad, and 250-600μm long, 2-3μm broad at the upper region, thick walled upto certain region and later thin wall, hyaline to sub-hyaline in the terminal region branches inverticils long or short with many orders like primary, secondary or tertiary. Conidiogenous cells phialidic, inverticils long 2-4μm at base, slowly narrower at the apex, 10-40μm long and 1-2μm broad at the apex with or without collarate. Conidia phialidic, one celled, smooth, cylindrical or ellipsoidal, 3-5μm long and 1-2μm broad.

From decaying plant debris from Warangal, Nizamabad, Bhadrachalam, Yellandu, Ashwaraopet forests of Khammam district, A.P. HCIO No:43, 046. NCBL No. 97, 231. Nov. 2002. Coll. V.R. & G.S.K.

Verticillium is well known through out the world for its pathogenic activity. It is known Andhra Pradesh from soil of Vikaradad and other places.

57. WIESNERIOMYCES Koorders 1907

Ver. Akad. Wet. Amst. 13 : 246.

The genus *Wiesneriomyces* was erected by Koorders (1907), with *W. javanicus* as the type species. Kirk (1984), examined the holotype of *Volutellaria laurina* Tassi and showed that this provided an earlier name for *Wiesneriomyces javanicus* and hence proposed the new combination *Wiesneriomyces caurinus* (Tassi) P.M. Kirk. *Volutellaria, Chaetopeltis laurina* (Tassi) Sacc., *Tassia Laurina* (Tassi) H& P. Sydow and *Wiesneriomyces javanicus* Koorders were all reduced by Kirk (1984) to synonymy with *W. laurinus*. The taxon is characterised by pulvinate sporodochia, sharp projecting setae, closely adpressed, branched, hyaline conidiophores, polyblastic conidiogenous cells, conidia filform and aggregated in slimy golden yellow masses.

The genus is evidently wide spread but remained monotypic and has been reported from many parts of the world including Java (Koorders, 1907),

India (Subramanian, 1956), (Matsushima, 1971), Japan (Matsushima 1975) Taiwan (Marsushima, 1980), the British Isles (Kirk,1983) and Australia. From India it has been reported from Madras (Subramanian, 1956), Maharastra (Patil, 1965) Madhya Pradesh (Sharma, 1971), Orissa (K.A. Reddy, 1983). The genus *Wiesnerionyces* can be compared with *Isthmolongispora* Matsu., and *Speirsopsis* Tubaki (Sensu Carmichael et al 1980) in appearance of conidiogenous cells and conidia. Ellis (1971)admits that it is probably one of the most widely distributed hyphomycetous genus in the tropical countries with some of the collection having smaller conidia which is an invaluable character in further speciation. Till recently it remained as monotypic genus, in 1988 Kuthubutheen & Nawavi from Malaysia reported the second species viz., *W. conjuctosporus* Kuthubutheen & Nawavi. In the present studies on some hyphomycetes of Nalgonda the material collected resemble the type species and hence the same is being described here.

Wiesneriomyces laurinus (Tassi) Kirk. Trans. Br. Mycol. Soc. 82: 748 - 749 1984. (Fig. 65)

Fig. 65. *Weisnerionyces laurinus*
a-b, Conidiometa with setae, conidiophores and conidia (450 x, 1000 x)

Colonies effuse, consisting of scattered pulvinate sporodochia, mycelium immersed. Sporodochia dark brown in colour, having stroma at the base from which hyaline to sub-hyaline conidiophores arise. These conidiophores in their turn bear light yellow mass of conidia encircled by a number of curved dark brown setae. Stroma erumpent or superficial brown, present at the base of sporodochia. Setae simple, long, inwardly curved , subulate, swollen at the base, acutely pointed at the apex, septate, dark brown, smooth, 500μm long, 9μm thick at the base. Conidiophores macronematous, close to one another and form sporodochia, narrow, branched at the apex, smooth, hyaline, straight or flexuous, 40μm long, 3μm wide. Conidiogenous cells develop at the end of the branches. Usually 3 in number, polyblastic, discrete, determinate, clavate or cylindrical , denticulate bearing chains of conidia. Conidia fromed in acropetal chains but becoming aggregated in slimy golden yellow masses. Individually colourless, smooth, 0-septate, the conidium at each end of a chain tapered intermediate ones more or less cylindrical. The conidia remain attached to one another for a long time by narrow isthmi or connectives and succeed with difficulty. 12-15 Conidia in each chain measuring 60-80μm long, each conidium 10-12μm long, 3-4μm wide.

Material examined on unidentified leaves near water bodies, Dindi, Nalgonda, Khammam, Vijayawada, Kakinada, A.P., NCBL No. 272, 345. Nov. 2002. Coll. V.R. & G.S.K.

58. XYLARIA Hill & Schrank

ex Dennis, Kew Bull. 401, 1957

Xylaria belongs to *Xylariaceae* Tul. & C. Tul., which is one of three families in order xylatariales with 46 genera and 795 species of Ascomycotina. Muller and Von Arx (1973), considered *Xylaria* of Xylariaceae in Sphaeriales under Ascomycotina. Cooke and Hawksworth (1970), recognised Xylariaceae L & *C* Tulsane (1861), family with *Xylaria* Hill & Grev. as its holotype. *Xylaria* produces more or less erect or stipitate stromata stalks sometimes embedded in the substrate. It produces stroma on plants or debris originating from termitaria, erect clavate, simple of branched with fertile head apex. Asci cylindrical with fertile (?) apex and producing 8 ascospores one celled, blackish brown smooth.

Xylaria hypoxylon Hill ex Grev.Fl. Edinb. Pp. 355, 1824 (Fig. 66)

Fig. 66. *Xylaria hypoxylon*
a-b, Ascomata, asci, c-e, V.S. of ascocarp showing ascospores (1000 x).

Xylaria produces cylindrical or elongated ascomata, bluish ash coloured, whitish brown colonies. Mycelium submerged in the substrate septate, branched, 2-3μm broad, sub-hyaline to brown, smooth thin or thick walled, smooth texura ascomata angularis or porreta. Ascomata cylindrical or irregular 2-4μm long, 0.5-0.7μm broad on crushing releasing the asci and paraphyses. Paraphyses thin walled septate, smooth, 2-3μm broad base, 3-4μm broad at apex, smooth apex specialised looking fertile. Ascospores one celled, obliquely placed in row one upon another, oval, smooth, 2-3μm in diam.

From Manoharabad, Ibrahimpatnam, Bhadrachalam, Kotagudem and Ashwaraopet forests of Khammam, Nalgonda, Medak districts of A.P., HCIO No: 43,047. NCBL No. 291, 350. Nov./Dec. 2002. Coll. V.R. & G.S.K.

IV List of Fungi Collected From Districts of Andhra Pradesh

IV List of Fungi in Districst of A.P.
(Published and to be Published)

ADILABAD

1. *Acrogenospora sporocephala* (Berk & Br.) Ellis
2. *Acrophragmis resingerii* Vasant Rao & Reddy
3. *Anellophora solani* (H. Syd.) Hughes
4. *Beltrania rhombica* Penzig
5. *Beltraniella odinae* Subramaniam
6. *Calostilbella indica* sp. nov.
7. *Catenularia hughesii* sp. nov.
8. *Centrophiale elagans* Gen. nov.
9. *Curvularia andropogonis* (Zimm.) Boedijn
10. *Dendryphiosphaera minuta* Vasant Rao & de Hoog
11. *Exosporium monanthotaxis* Pirozyski
12. *Geniculisporium serpens* Chesters & Greenhalgh
13. *Gliocephalotrichum bulbilium* Ellis & Hesseltine
14. *Graphium pencilloides* Corda
15. *Hansfordia ovalispora* Hughes
16. *Haplographium mangiferae* Deighton
17. *Helicoma ambiens* Morgan
18. *Helicoma simplex* Rao & Rao
19. *Helicosporium gracile* (Morgan) Linder
20. *Helicosporium hiospiroides* Rao et al
21. *Heliocephala prolifarans* Vasant Rao et al
22. *Hermatomyces tucumanensis* Spegazzini
23. *Lappodochium laginiforme* Matsushima

24. *Linodochium hyalinum* Hoehnel
25. *Montosporella rhizoidea* Vasant Rao & de Hoog
26. *Papulaspora sepedonioides* Preuss
27. *Periconia atropurpurea* (Berk. & Curt), Litvinov
28. *Periconiella cocoes* Ellis
29. *Pestalotia pezizoides* De Not
30. *Petrakia echinata* (Peglion), Sydow
31. *Phaeoisaria clematidis* (Fuekel) Hughes
32. *Pithomyces ellisii* Vasant Rao & Chary
33. *Pleurophragmium tritici* Ellis
34. *Pycnostycanus azaleae* (Peck.) Mason
35. *Ramaraomyces corticola* Rao, Chary & Goos
36. *Scolecobasidium variable* Barron & Busch
37. *Spadicoides indicus* sp. nov.
38. *Sporochisma nigroseptata* Rao & Rao
39. *Tetraploa aristata* Berk & Br.
40. *Torula herbarum* Link & Gray
41. *Trichoderma viride* Pers.ex Fries
42. *Verticillium speciosa* sp. nov.
43. *Wiesnariomyces javanicus* Kooders

NIZAMABAD

1. *Acrodictys globulosa* Ellis
2. *Actinocladium rhodosporum* Ehrenberg ex Pers.
3. *Bahusandhika sundara* Vasant Rao & Dev Rao
4. *Beltrania rhombica* Penzig
5. *Brachysdesmiella biseptata* Vasant Rao & de Hoog
6. *Brachysporiella bulbosa* Rao & Rao
7. *Catenularia* state of *Cheatosphaeria novae-zelandiae* Hughes et Shoemaker
8. *Cladosporium herbarum* (Pers.) Link ex SF Gray
9. *Cheiropolyschema formasana* Matsushima
10. *Dictyoarthrinium sacchari* (Stev.) Damon
11. *Exserticlava vasiformis* (Matsu.) Hughes
12. *Eversia indica* sp.nov.
13. *Gyrothrix podosperma* (Corda) Rabenhorst.
14. *Helicoma* sp.1
    H.elicornea sp. 2
15. *Kumausanuha* sp.
16. *Melanocephala triseptata* Hughes
17. *Moorella speciosa* Rao & Rao
18. *Pithonyces charterum* (Berk. & Por.) Ellis
19. *Petrakia echinata* (Peglion) Syd.
20. *Periconia byssoides* Pers. ex Merat
    P.hispidula (Pers. ex Pers.) Mason & Ellis
21. *Sporidesmium ghanaense* Ellis
    S.longisporum Sharma & Munjal.
22. *Sporoschisma nigroseptata* Rao & Rao
23. *Stachybotrys* sp. nov.
24. *Syncephalastrum* sp.

25. *Spadicoides atra* Hughes
26. *Tritospeira ugadansis* Pirozynski
27. *Torula herbarum* Link & Gray
28. *Trimmatostroma scutellave* Ellis.
29. *Tharoopama trina* Subramanian
30. *Volutella ciliata* (Albertini & Sehwentrz) Fries
31. *Wiesneriomyces laurinus* Kirk 1984
32. *Xylohypha nigricans* Mason

MEDAK

1. *Actinocladium agumbense* sp. nov.
   A. rhodosporum Ehrenberg
2. *Agaricostilbum indicum* Subram. & Nat. comb. nov.
3. *Alsydiopsis agumbensis.* sp.nov
4. *Ampulliferina aseptoconidiate* sp. nov.
5. *Bactrodesmium longisporum* Ellis
6. *Beltrania rhombica* Penzig
7. *Beltraniella aethiopica* Kendrick & Bhat
8. *Berkleasmium indicum* sp. nov.
   B. pulchrum Holubova-Jechava & Mercado
9. *Brachydesmiella biseptata* var. *orientalis* Vasanth Rao & de Hoog
10. *Brachysporiella gayana* Batista
    B. setosa Grove et Saccardo
11. *Catenularia* sp.
12. *Chalara pteridina* Sydow
13. *Cheiromyces recurvous* Vasant Rao & de Hoog
    C. stellatus Berkely & Curtis
14. *Cicatrospira* Gen.et sp. nov.
    C. indica sp. nov.
15. *Clasterosporium bhagavatianum* sp. nov.
16. *Consetiella indica* sp. nov.
17. *Cryptophiale kakombensis* Pirozynski
18. *Cryptophialoidea secunda* Kuthubutheen & Nawawi
19. *Dactylaria furscispora* sp. nov.
20. *Dactylella benbicoides* Dreschler
21. *Dendryphiopsis atra* Hughes

22. *Dictyosporium heptasporum* Damon
23. *Domingoella asterinarum* Petrak & Ciferri
24. *Edumundmasonia pulchra* Subramanian
25. *Eduocalyx thwaitesii* Berkeley & Broome
26. *Endophragmiopsis pirozynski* Ellis
27. *Excipulariopsis narsapurensis.* (Subram.) Kirk & Spooner
28. *Exerticlava globosa* Vasant Rao & de Hoog

    E. vasiformis Hughes
29. *Gonytrichum macrocladium* Hughes
30. *Gyrothrix cornuta* Vasant Rao & de Hoog

    G. podosperma Rabenhorst
31. *Janetia indica* sp. nov.
32. *Kaybediomycetiella* Gen. et sp. nov.
33. *Kionochaeta ramifera* Kirk & Sutton

    K. setosa sp. nov
34. *Kionochaetella* Gen. et sp. nov.

    K. indica sp. nov.
35. *Meliola* species
36. *Monodictys abuensis* Vasant Rao & de Hoog

    M. putridense Hughes
37. *Monotosporella rhizoidea* Vasant Rao & de Hoog
38. *Moorella speciosa* Rao & Rao
39. *Mycoenterolobium platysporum* Goos
40. *Nodulosporium tuberum* Fontana & Bonfante ex de Hoog
41. *Phialocephala xalopensis* Maggi & Persiani
42. *Pithomyces ellisii* Vasant Rao & Chary

    P. subramanianii Vasant Rao & Reddy
43. *Pseudoustilaginodiella* Gen. et sp. nov.

    P. agumbica sp. nov.

44. *Pyriculariopsis indica* sp. nov.
45. *Spadicoides atra* Hughes
    S. obovata Hughes
46. *Sporendocladia foliicola* M. Wingfield
47. *Sporidesmium circinocandatum* sp. nov.
    S. garryae Bonar
    S. leonense Ellis
    S. vagum Nees & Nees
48. *Stachybotrys* sp.
49. *Stigmina shoreansis* sp. nov.
50. *Syncephalastrum racemosum* Cohn ex Schroter
51. *Taeniolella rubra* sp. nov.
52. *Tetraploa aristata* Berkely & Broome
53. *Trichobotrys effusa* Berkely & Broome
54. *Trimmatostroma scutellare* Ellis
55. *Uberispora simplex* Pirozynski & Hodges
56. *Veronea musae* Ellis
57. *Virgaria nigra* Nees ex SF Gray
58. *Virgariella atra* Hughes
59. *Volutella ciliata* Fries
60. *Wiesneriomyces laurinus* Krik
61. *Zygosporium gibbum* Hughes
    Z. masonii Hughes
    Z. oscheoides Montangne

NALGONDA

1. *Acladium ramosissium* Ellis
2. *Acrogenospora sporocephala* (Bark. & Br.) Ellis
3. *Arbuscula euginiae* Batista & Pers.
4. *Beltrania rhombica* Penzig
5. *Beltraniella odinae* Subramanian
6. *Codineae simplex* Hughes & Kendrick
7. *Chalara fusidioides* (Corda) Rabenh
8. *Curvularia lunata* (Walkker) *Boedijn*
   C. tuberculata Jain
9. *Dictyoarthrinium sacchari* (Stev.) Damon
10. *Dreschlera* sp.
11. *Exserticlava vasiformis* (Matsu.) Hughes
12. *Gliomastrix murorum* (Corda) Hughes
13. *Gyrothrix cornuta* Vasant Rao & de Hoog
14. *Helicosporium vegetam* Nees
    H. hiospiroides Rao et al
15. *Linodochium hyalinum* (Lib.) Hoehnel
16. *Melanospora zaminae* Corda
17. *Myrothecium bisetosum* Vasant Rao & de Hoog
18. *Perriconia byssoides* Pers. ex Merat
    P. hispidula Mason & Ellis
    P. kookei Mason & Ellis
19. *Pencillium chrysogenum* Thom.
20. *Pithomyces chrtarum* Berk. & Br.
21. *Puccinia* sp.
22. *Rhizoctonia* sp.
23. *Spadicoides obovata* Hughes

24. *Sporidesmium garryae* Bonar

    S. vagum Nees & Nees

25. *Stachybotrys globosa* Misra & Srivastava
26. *Syncephalastrum racemosum* Cohn ex Schroter
27. *Tetraploa aristata* Berk. & Br.
28. *Totrula herbarum* Link & SF Gray
29. *Trimmatostroma scutellara* Ellis
30. *Tharoopama trina* Subramanian
31. *Trichoderma viride* Pers. ex Fries

    T. koningii Oundeman

32. *Virgaria nigra* Nees ex SF Grey
33. *Volutella ciliata* Fries
34. *Verticillium theobromae* Mason & Hughes
35. *Zygosporium echinosporum* Bunting & Mason

KARIMNAGAR

1. *Acladium ramosissium* Ellis
2. *Acrodictys globulosa* Ellis
3. *Ampuliferina indica* sp.nov.
4. *Bahugada sundara* Vasant Rao & Reddy
5. *Bahusandhika sundara* Vasant Rao & Dev Rao
6. *Beltrania rhombica* Penzig
7. *Bilgramia mahadevpoorensis* sp. nov.
8. *Brachysporiella bulbosa* Reddy

   B. gayana Batista
9. *Circinotrichum mahadevpoorensis* sp.nov.
10. *Codinaea simplex* Hughes & Kendrick
11. *Cordana ellipsoidea* de Hoog
12. *Corynespora* sp.nov.
13. *Cryptophiala guadakanalense* Matsushima
14. *Cylindrocladium scoparium* Morgan
15. *Deightoniella torulosa* Ellis
16. *Dendryphiopsis atra* Hughes
17. *Dictyoarthrinium sacchari* (Stev.) Damon
18. *Digitodesium hexasporum* Reddy
19. *Diplococcium spicatum* Groove
20. *Eversia indica* sp.nov.
21. *Exserticlava vasiformis* (Matsu.) Hughes
22. *Farlowiella acrogenospora* conidial state Saccardo
23. *Helicoma palmigenum* (Penzig & Saccardo) Linder

24. *Hypocrea trichoderma* conidial state Dingley
25. *Kumanasamuha sundara* Rao & Rao
26. *Mahadevania speciosa* sp. nov.
27. *Melanocephala triseptata* Hughes
28. *Melanographium citri* Ellis
29. *Mycoenterolobium platysporum* Goos
30. *Nectria verticillium* conidial state Fries
31. *Periconia byssoides* Pers. ex Marat

    P. hispidula (Pers. ex Pers.) Mason & Ellis

    P. pseudolateralis de Hoog & Vasant Rao

    P. minutissima Corda

    P. ramostipitata Patwardhan & Desai
32. *Piricauda pseudarthriae* Ellis
33. *Pseudobotrytis bisbyi* Timonin
34. *Shahalamomyces indicus* Vasant Rao & Khan
35. *Sporidesmiella cuneiformis* (Sutton) Kirk
36. *Sporidesmium ghanaense* Ellis

    S. longisporum Sharma & Munjal

    S. macrunum Ellis
37. *Sporoschisma nigroseptata* Rao & Rao
38. *Stemonitis unifera* Mabr.
39. *Stigmina compacta* (Sacc.) Ellis
40. *Tharooopama trina* Subramanian
41. *Torula herberum* Link & Gray
42. *Tretopileus sphaerophorus* Hughes & Deighton
43. *Trichothecium roseum* (Pres.) Link ex Fries
44. *Trimmatostrama scutellare* (Berk. & Br.) Ellis

45. *Tritirachium dependens* Limber
46. *Virgaria nigra* (Link) Nees ex Gray
47. *Virgariella atra* Hughes
48. *Virgatospora echinofibrosa* Finley
49. *Xylohypha nigrescens* (Pers. ex Fr.) Mason

KHAMMAM

1. *Acrospeira nensis sp.*
2. *Acroconidiella arecae* Ellis
3. *Annellodochium khammamatum* sp. nov.
4. *Arbuscula euginae* Batista & Pers.
5. *Bahusakala olivaceonigra* Subramanian
6. *Beltrania rhombica* Penzig
7. *Beltraniella odinae* Subramanian
8. *Caeamchariomyces* Gen. nov.
   C. liniodocheana sp. nov.
9. *Cancellidium applanatum* Tubaki
10. *Chaetochalara africana* Sutton & Pirozynski
11. *Chaetomella circinoseta* Stolk
12. *Chaetomium bostrychodes* Zonf.
11. *Chaetopsina splendida* Sutton & Hodges
12. *Chalara alabamensis* Morgan - Jones & Ingram
    C. fusidioides (Corda) Rabenlh
13. *Chintooromyces* Gen. nov.
    C. elagans sp. nov.
14. *Chintooromycetella* Gen. nov.
    C. majas sp. nov.
15. *Circinotrichum fertile* Pirozynski & Hodges
    C. poonenais Pinozynski & Pattil
16. *Codinaea septata* Sutton & Hodges
17. *Cylindrocarpon aquaticum* Comb.nov. Mervanova & Descals
18. *Dictyoarthrenium sacchari* (Stev.) Damon
19. *Dityochaeta abnormis* Holubova - Jachova
    D. querna Kirk

20. *Ellisiopsis occulata* Rambelli & Lunghini
21. *Fusariella atroviens* Saccardo
22. *Gliocladium pencillioides* Corda
23. *Gyrothrix hughesii* Pirozynski
    G. cornuta Vasant Rao & de Hoog
24. *Hansfordia carcis* Krik
25. *Heliocephala proliferans* Vasant Rao et al
26. *Helicosporium vegetum* Nees
27. *Hemibeltrania mitrata* Kirk
28. *Lacellina graminicola* (Berk. & Br.) Patch
29. *Lacelliniopsis sacchari* Subramanian
30. *Mallariomyces* Gen. nov.
    M. elagans sp. nov.
31. *Melanospora zaminae* Corda
32. *Menisporopsis theobromae* Hughes
    M. pleiosetosa Vasant Rao & de Hoog
33. *Myrothecium bisetosum* Vasant Rao et al
34. *Nodulisporium tuberum* Fontana & Bonfante ex de Hoog
35. *Parapleurothesiopsis* Kirk
    P. indicus sp. nov.
36. *Parapseudofusidium* Gen.nov.
    P. indicus sp. nov.
37. *Periconia hispidula* Mason & Ellis
    P. lateralis Ellis & Everhart
    P. laminella Mason & Ellis
    P. byssoides Pers. ex Merat
38. *Phalangiospora constricta* Nawawi & Webster

39. *Pithomyces chartarum* (Berk. & Curt.) Ellis
40. *Puccinia* species
41. *Radhakrishnomyees* Gen.nov.

    R. radiatus sp. nov.
42. *Serquicillium setosum* Vittal

    S. canadelabrum (Bonorden) Gams
43. *Spadicoides aggregata* Subramanian & Vittal
44. *Spegazzinia tessarthra* (Berk & Curt.) Saccardo
45. *Stachybotrys globosa* Misra & Srivastava

    S. pulchara Speg.
46. *Stemonitis flavogenita* John
47. *Subbarajuensis* Gen. nov.

    S. indica sp. nov.
48. *Subramaniomyces fusissaprophyte* (Matsu.) Kirk
49. *Torula herbarum* Link ex Gray
50. *Trichoderma viride* Pers. ex Fries
51. *Volutella ramkumari* Sarbhoy
52. *Wiesneriomyces laurinus* Krik
53. *Zygosporium masoni* Hughes

    Z. echinosporum Bunting & Mason

SRIKAKULAM

1. *Allescheriella indica* Vasant Rao
2. *Alterneria brassicicola* (Schw.) Wiltshire
   A. radicina Meiev. Dreschler & Eddy
   A. saparva (Subram.) Deighton
3. *Angulimaya sundara* Subramanian & Lodha
4. *Aspergillus fumigatus* Freeseuicus
   A. flavus Link ex Fries
   A. nidulens (Eidam) Winter
   A. versicolor (Vuillemin) Tiraboeschil
   A. candidus Link ex Fries
5. *Bactrodesmium atrum* Ellis
6. *Beltrania rhombica* Penzig
7. *Circinotrichum maculiforme* C G Nees ex Fers.
8. *Cryptophiale kakombensis* Pirozynski
9. *Conoplea mandasa* sp. nov.
10. *Curvularia prasadii* Marthur RL & BL Chona
    C. lunata (Wakker) Boedijn
    C. stapeliae Hughes & da Plessis
    C. tuberculata Jain
11. *Dendryphiopsis atra* Hughes
12. *Dictyoarthrinium sacchari* (Stev.) Damon
13. *Dictyosporium mandasii* sp.nov
14. *Dreschelra* sp.
15. *Excipulariopsis narsapurensis* Subramanian
16. *Exosporium tiliae* Link

17. *Gliomastix murorum* (Corda) Hughes
18. *Graphium putredinis* (Corda) Hughes

    G. pencilloides Corda
19. *Idriella vandalurensis* Vittal & Dorai
20. *Lacellina graminicola* (Berk. & Br.) Petch
21. *Latericonis obscura* Vasant Rao et al
22. *Memnoniella echinata* (Riv.) Galloway
23. *Monodictys putredinis* (Wallr.) Hughes
24. *Olpitrichum patulum* Holubova - Jachova
25. *Ophiostoma* sp.
26. *Papulaspora sepedonioides* Preuss
27. *Paradictyoarthrium differuctum* Matsushima
28. *Phaeoisaria* sp.
29. *Pencillium chrysogenum* Thom.

    P. steckii Zaleski
30. *Periconia byssoides* Pers. ex Schwein.

    P. hispidula (Pers. ex Paers) Mason & Ellis

    P. tirupatiensis Subramanian
31. *Periconiella cocoes* Ellis
32. *Pestalotia pezizoides* De Not
33. *Pithomyces chartarum* (Berk Br.) Ellis
34. *Rhizoctonia* sp.
35. *Scolecobasidium variable* Barron & Busch
36. *Spegazzinia intermidia* Ellis
37. *Spiropes* sp.
38. *Srikakulomyces donubai* Gen. et sp. nov.
39. *Syncephalastrum racemosum* Cohn ex Schroter

40. *Taeniolella rubra* Hughes
41. *Tharoopama trina* Subramanian
42. *Torula harberum* Pers. ex SF Gray
43. *Trichothecium roseum* (Pres.) Link ex Fries
44. *Tritirachium dependens* Limber
45. *Xenosporium indicum* Panwar, Prohit & Gehlot.

V References

Abott, E.V., 1927 *Scolecobasidium* a new genus of Soil Fungi, Mycologia, 19:29-31.

Agarwal, G.P., & S.K., Hasija, 1961, Fungi causing plant diseases in Jabalpur-VIII J.Indian.bot.Soc., 40: 542-547.

Agnihothrudu, V., 1961, A list of Indian Myxomycetes., J. Madras Univ. B. Madras, Pp. 1-20

Alexopoulos, C.J., 1978, Myxomycetes in Taxonomy of Fungi-(ed. Subramanian) Univ. of Madras, India.

Alexopoulos, C.J., C.W. Mims & M. Black Well., 1996 Introductory Mycology Johnwilly & Sons, New York, Toronto.

Ainsworth, G.C. & A.S. Sussman (eds.), 1965, The Fungi-An Advanced Treatise, Vol.I. Acad. Pr. New York, London.

Ainswroth, G.C., F.K. Sparrow & A.S. Sussman (eds.), 1968, The Fungi-An Advanced Treatise, Vol.III. Acad. Pr. New York, London.

Ainsworth, G.C., F.K. Sparrow & A.S. Sussman (eds.), 1973, The Fungi-An Advanced Treatise, Vols.IV.A.& IV.B. Acad.Pr. New York, London.

Atkinson, G.F., 1984 *Olpitrichum* a new genus of Mucedinous fungi, Bot.Gaz.19: 244-247.

Barron, G.L., 1964, A note on the relationship between *Stachybotrys* and *Hyalostachybotrys.* Mycologia, 56:313-316.

Barron, G.L., 1968, The Genera of Hyphomycetes from soil. Williams & Wilkins, Baltimore Maryland.Pp. 1-364.

Batista, A.C., 1952, Dois novos Fungus Imperfectos Bol. Agric. Pernambuco 19:108-111.

Batista, A.C. & R.Ciferi. 1963, The sootymolds of the faimly Asboliciaceae. Quaderno 31:1-229.

Bhat, D.J. & W.B. Kendrick. 1993, Twenty five new conidial fungi from the Western Ghats and The Andaman Islands (India), Mycotaxon XLIX Pp. 19-90.

Bhattacharia, B. & H.K. Baruah. 1953, Fungi of Assam, J. Univ. Gauhati, 4:287-312

Bilgrami, K.S., J.Jamaluddin & M.A. Rizwi. 1991, Fungi of India Part-I. Today & Tomorrow's Pr. Publ. New Delhi. Pp. 1-467.

Booth, C. 1971 a, Introduction to General Methods in Methods in Microbiology ed. Booth, C Acad.Pr.New York. Pp.79-110.

Booth, C. 1971 b, Fungal Culture media In Methods in Microbiology ed. Booth, C. Acad. Pr. New York. Pp. 1-94.

Bisby, G.R., 1943, *Stachybotrys.* Trans. Brit. mycol. Soc. 26:133-143.

Bisby, G.R., 1945 *Stachybotrys* and *Memnoniella.* Trans. Brit. mycol. Soc. 28:11-12.

Carlile, M.J., S.C. Watkinson & G.W. Gooday, 2001 The Fungi. Acad. Pr. Harcourt Brace & Co.Pub.London, Boston.Pp.1-594.

Carmichael, J.W., W.B. Kendrick, I.L. Conners, and L. Sigler. 1980, Genera of Hyphomycetes. The Univ. of Alberta Pr. Alberta, Canada. Pp. 1-386.s

Castaneda Ruiz., R.F. 1986, Deuteromycotina de Cuba Hyphomycetes-IV ed. en Instituto de Investigaciones Fundamentales en Agricultura Tropical. Alejandro de Humboldt. Acad. de ciencias de Cuba, la Habana Pp. 17.s

Castaner, D., 1968 *Monacrosporium obtrulloides* a new hyphomycete capturing nematodes in constricting wings. Can. J. Bot. 46: 763-765.s

Cesati, V. & G. De Notaris., 1863 Schema di classifictione Persoon.Com.Soc.Brit.Ital. 1:177-240 (Microfilm).

Chang, S.T.& Miles, P.G., 1993 Edible Mushrooms and their Cultivation, CBS Publ.& Dist. New Delhi. Pp.1-345.

Child, M., 1932 The Genus *Daldinia,* Ann. Mo.Brit. Gdn. 19: 429-496.

Clements, F.E. & C.L. Shear, 1931 The Genera of Fungi, Wilson, New York. pp. 1-496.

Cole, G.T., 1974 Conidiophore and conidium ontogeny in *Spegazzinia tessarthra* Can. J. Bot. 52:1259-1264.

Cole, G.T. & R.A. Samson, 1979 Patterns of development in Conidial Fungi, Pitman London. Pp. 1-190.

Cole, G.T. & W.B. Kendrick, 1981 Biology of conidial Fungi in Vol. I & II Pp. 1-486, & 1-660. Acad. Pr. New York, Toronto, London.

Cooke, M.C., 1881 Illustrations of British Fungi (Hyphomycetes), London in 8 Vols. (4 Microfiche).

Cooke, R., 1981 Fungi, William Collins & Sons, London. Pp. 1-159.

Cooke, R.C., 1967 *Monacrosporium globosporum* and other net work forming Nematode Trapping species. Trans. Brit. Mycol. Soc. 50: 515-518.

Cooke, W.B. & D.L. Hawksworth, 1970 A Preliminary List of the families Proposed for fungi (including the Lichens) Mycol. Pap. No:121 CMI 86 Pp.

Corda, A.C.J., 1829-1841 Die Pilze Deutschlands und Bohmens in Sturm's Floora II,III.

Corda, A.C.J., 1837-1854 Icones Fungorum hucuspue Cognitorum Vol. I-VI, Prague.

Cunningham, J.L., 1974 A new *Gyrothrix* in culture and a Key to its species. Mycologia 66: 122-129.

Damon, S.C., 1953 Notes on Hyphomycetous genera, Bull. Toerrey bot.Cl. 80: 155-165.

de Hoog, G.S., 1972 The genera *Beauveria, Isaria, Tritirachium* and *Acrodontium* Gen.nov. stud. Mycol. No.1: 1-41.

de Hoog,G.S.&J.A. von Arx., 1973 Revision of *Scolecobasidum* and *Pleurophragmium* Kavaka 1:55-60.s

de Hoog,G.S. & E.J. Hermanides, Nijhoff., 1977 Survey of the black Yeasts and allied fungi. Stud. Mycol. 15: 1-140.

Deighton, F.C.& K.A. Pirozynski, 1972 Microfungi-V more Hyperparasitic Hyphomycetes. Mycol. Pap. 128: 1-110.

Dennis, R.W.G., 1963 Hypoxyloideae of Congo, Bull. Jard. Bot. Etat. Brux 33: 317-343.

Dennis, R.W.G., 1968 British Ascomycetes J. Cramer, Lehre. P.455.

Dickinson, D.H., 1968 *Gliomastix,* Mycol.Pap.No. 115: 1-24.

Domsch, K.H., W. Gams & T.H. Anderson, 1980 Compendium of Soli Fungi in 2 Vols. Acad. Pr. New York, London.

Donk, M.A., 1966 Check List of European hymenomycetous Heterobasidiae, Persoonia 4: 145-335.

Dorai, M. & B.P.R. Vittal., 1986 A new species from Eucalyptus litter. Trans. Brit. mycol. Soc. 87(4): 642-644.

Drechsler, C., 1937 Some Hyphomycetes that prey on free living terricolous nematodes. Mycologia 29: 446-552.

Drechsler, C., 1952 Another nematode strangulating *Dactylella* and some related Hyphomycetes. Mycologia 44: 533-556.

Drechsler, C., 1962 Some clampless Hyphomycetes predacious on Nematodes and Rhizopods. Sydowia 15: 9-25. (1961).

Dring, D.M., 1973 Gasteromycetes (in The Fungi. Ainsworth et al.) Acad.Pr. New York London.Pp.451-478.

Dwivedi, R.S., 1959 Occurrence of the genus *Scolecobasidium* Abbott in India. Curr.Sci.28: 374-375.

Ellis, M.B., 1949 *Tetraploa* Trans.Brit.mycol. Soc. 32: 246-251.

Ellis, M.B., 1961 Dematiaceous Hyphomycetes - II Mycol. Pap. 79: 1-23.

Ellis, M.B., 1961 Dematiaceious Hyphomycetes - III Mycol. Pap. 82: 1-55.

Ellis, M.B., 1967 Dematiaceious Hyphomycetes - V Mycol. Pap. 111: 1-46.

Ellis, M.B., 1971 Dematiaceious Hyphomycetes CAB, Comm. Inst. Kew, England. Pp. 608.

Ellis, M.B., 1972 Dematiaceious Hyphomycetes - XI Mycol.Pap. 131: 1-25.

Ellis, M.B., 1976 More Dematiaceious Hyphomycetes CAB, CMI, Kew, England, Pp. 1-507.

Ellis, M.B. & J.P. Ellis., 1985 Microfungi on land plants. Croom Helm. London & Sydney Pp. 1-818.

Fries, E.M., 1821-1982 Systema Mycologium, systems, fungorum ordines, genera et species hucusque cognitas 3 Vols. Pp. 1866. Greifswald.

Fries, E.M., 1846 - 1849 Sumna vegetabilium Scandinaviae, Hollmia et Leipzig.

Fries, S.M., 1989 Stink horns. The mycologist, IMI, 3 (4): 190-191.

Gams, W., 1971 *Cephalosporium artige* Schimmel plize Gustav. Fischer, Verlag, Stutgart, Pp. 1-261.

Gams, W., 1974 Generic concepts in Hyphomycetes with slimy 1-celled phialoconidia in taxonomy of Fungi (Subramanian, C.V. ed.) Univ. of Madras, madras.

Gams, W. & V. Holubova-Jechova., 1976 *Chloridium* and some other phialidic dematiaceous genera growing on decaying Stud. Mycol. 13:1-99.

Gangulee, H.C. & A,.K. Kar., 1985 College Botany. Vol. II Pp. 219-524. New Central Book Agency, Calcutta.

Gaumann, E.A., 1964 Die Pilze, 2nd ed. Birknaeuser, Basel.

Gaumann, E.A. & C.W.D. Dodge., 1928 Comparative morphology of Fungi, Mc Graw Hill, New York.

Grainger. J., 1962 Vegetative and fructifying growth in *Phallus impudicus* Trans. Brit. mycol. Soc. 45: 147-155

Guba, E.F., 1961 Monograph of *Monochaetia* and *Pestalotia.* Harward Univ. Pr. Cambridge., Mass

Guzman, G., 1970 Monographia del genero *Scleroderma* Darwinia 16: 233-407.

Hashmi, M.H., 1973 Conidium ontogeny in Hyphomycetes *Arthrinium* and *Dictyoarthrinium,* Pak. J. Bot. 5: 119-124.

Hawksworth, D.L., 1974 Mycologists Hand book. M.I. Kew, Surrey, England. Pp. 1-229.

Hawkshwroth, D.L., P.M. Kirk., B.C. Sutton & D.N. Pegler., 1995 Ainsworth & Bisby's Dictionary of the Fungi. CAB International, Walligford, Oxon, U.K.

Hohnel F. Von., 1909 Fragmente Zur Mykologie. Sber Akad. Wiss. Wien. Pp 118: 275-1552

Hohnel F. Von., 1923 System der Fungi Imperfecti Fuckel. Mycol. Unters. 3:301-369.

Hughes, S.J., 1951 Studies on Mcrofungi - IX, Mycol. Pap. 43: 1-25.

Hughes, S.J., 1952 Fungi from Gold coast-I, Mycol. pap. 48: 1-91.

Hughes, S.J., 1952, Studies on Microfungi-XIII Mycol. Pap. 47: 1-15.

Hughes, S.J., 1953 Conidiophores, Conidia and classification Can. J. Bot. 31:577-659

Hughes, S.J., 1958 Revisiones hyphomycetum aliquot cum appendice de nominibus rejiciendis, Can. J. Bot. 36: 727-836.

Hughes, S.J., 1978 New Zealand Fungi - 25 miscellaneous species. New Zealand J. Bot. 16:311-370.

Hughes, S.J., 1979 Relocation of species of *Endophragmia* Auct, New Zealand J. Bot. 17: 139-188.

Ingram, D.S. & A. Hudson., 1995 Shape and form in plants and Fungi, Acad. Pr. for Linnean Society of London Ser. 16.

Jong, S.C. & E.E. Davis., 1976 Contributions to the knowledge of *Stachybotrys and Memnoniella* in culture Mycotaxon, 3:409-485.

Karsten. P.A., 1881 Enumeratio *Thalophorearum Fr. et Clavariarum Fr. Feunicarun, systematic novo dispositarium,* Rev. Mycol. 3:21-23.

Kendrick, W.B.(ed.), 1971 Taxonomy of Fungi Imperfeti. Univ. of Toronto Pr. Toronto. Pp. 309.

Kendrick, W.B., 1979 The whole Fungus Vols. I & II National Museum of Canada, Ottawa, Canada and the kananaskis Foundation. Pp. 1-793.

Kendrick, W.B., 1980 The generic concept in Hyphomycetes a reappraisal Mycotaxon, 11:339-364.

Kendrick, W.B. & R. Watling., 1979 Mitospores in Basidiomycetes. The whole Fungus Vol. 2:473-546. National Museums of Canada, Ottawa, Canada and the Kananaskis Foundation.

Kirk, P.M., 1983 New or interesting microfungi-XI Mycotaxon 18(2): 259-298.

Kirk, P.M., 1985 New or interesting microfungi-XIV Mycotaxon 23: 305-352.

Kirk, P.M., 1986 Sylloge Fungorum Omnium hucusque cognitorium Vol. 27.

Kirk, P.M. & B.C. Sutton., 1985 A reassessment of the anamorph genus *Chaetopsina* (Hyphomycetes). Trans. Brit. mycol. Soc. 85: 709-718.

Kohlmeyer, J., 1968 Marine fungi from the tropics, Mycologia, 60: 252-270.

Kohlmeyer, J. & E. Kohlmeyer., 1979 Marine Mycology, Acad. Pr. New York, London. Pp 1-690.

Kushwaha, R.K.S. & S.C. Agarwal., 1977 *Chrysosporium crassi* tunicatum sp.nov. a Keratinophitic fungus Trans. Brit. mycol. Soc. 68: 464-467.

Lakhanpal, T.N. & K.G. Mukerji., 1981 Taxonomy of the Indian Myxomycetes, J. Cramer, Vaduz. Pp. 1-530.

Lindau, G., 1900 Fungi Imperfecti in Engler & Prantil's Die Nat. Pflanz. famili. 1:347-523. Leipzing.

Lindau, G., 1904-1910 Hyphomycetes in Rabenhorst's Crypyo. Fl. 8: 1-852, 9: 1-983. Leipzig.

Lindau, G., 1922 Hyphomycetes in Die mikro. Pilze. Krypto. Fl. fur Arf. 2: 1-301. Berlin.

Linnaeus, C., 1753 Species Plantarum 2 Vols. Stockholm (Facsimile reprit, 1957-1959) Ray. Soc. London.

Liu, X.Z. & Zhang, K.Q., 1994 Nematode trapping species of *Monacrosporium* with special reference to two new species. Mycol. Res. 98: 862-868.

Lodha, B.C., 1970 A new species of *Scolecobasidium* Curr. Sci. 46 No: 22, 788.

Martin, G.W., 1952 Revision of the North Central Tremelales Stud. Nat. Hist. Iowa Univ. 19: 1-122.

Martin. P., 1967 Studies in the Xylariaceae I-IV Jl. S. Africa Bot. 33: 205-240, 315-328, 34: 153-199, 303-330.

Martin, P. & C.J. Alexopoulos., 1969 Morphology of the Myxomycetes. Iowa city, University of Iowa Pr. Pp. 560.

Mason, E.W., 1937 Anotated Account of Fungi received at the Imperial Mycological Institute, List II (Fascicle 3 genera part) Mycol. Pap. 4: 69-99.

Mason, E.W., 1941 Annoted Account of Fungi received at Imperial Mycological Institute, List-II (Fascicle 3 Special part) Mycol. Pap. 5: 101-144.

Mason, E.W. & M.B. Ellis., 1953 British species of *Periconia,* Mycol. Pap. 56: 1-127.

Mathur, R.S., 1979 The coelomycetes of India. Bishen Singh, Mahendrapal Singh. Dehradun, India. Pp. 1-460.

Matsushima, T., 1971 Microfungi of the Solomon Islands & Papua, New Guinea. Nippon printing & Publ. co. Osaka, Japan. Pp. 78.

Matsushima, T., 1975 Icones Microfungorum a Matsushima Lectorum, Kobe, Japan. Pp. 209.

Matsushima, T., 1980 Matsushima Mycological memoirs No.1 Pp.1-82.

Matsushima, T., 1981 Matsushima Mycological memoirs No. 2 Pp. 1-68.

Matsushima, T., 1983 Matsushima Mycological memoirs No. 3 Pp. 1-90.

Matsushima, T., 1985 Matsushima Mycological memoirs No. 4 Pp. 1-68.

Matsushima, T., 1987 Matsushima Mycological memoirs No. 5 Pp. 1-100.

Matsushima, T., 1996 Matsushima Mycological memoirs No. 9 Pp. 1-40.

Mereado Sierra, A., 1981 Lista Preliminar de hifomicetes demaciaceos de la Estacion Ecologica de Sierra del Rosario Y zonas adyacentes. Acta Bot. Cubana 6: 1-6.

Mereado Sierra, A., 1984 Hifomicetes demaciaceos de Sierra del Rosario, cuba. Editorial Academia La Habana, Cuba 181 Pp. 117, Laminas.

Mereado Sierra, A., V. Holubova-Jechova & J. Mena Portales., 1997 Hifomicetes demaciaceos de Cuba Enteroblasticos Maseo Reginale di Scienze Naturali; Torino. Pp 1-388.

Michaeli, P.A., 1729 Nova plantarum genera juxta Tournefortill Methodum disposita Pp. 234.

Minter, D.W., 1984 New concepts in the interpretation of conidiogenesis in deuteromycetes. Microbiological Science Vol. 1(4): 86-89.

Misra, P.C., 1967 *Torula terristris* new specis from soil, Can. J. Bot. 45: 367 - 369.

Moore, R.T., 1958 Deuteromycetes-I The *Sporidesmium* complex, Mycologia 50: 681-692.

Moore, R.T., 1959 The genus *Berkleasmium* Mycologia 51: 734-739.

Morgan-Jones, G. & R.C. Sinclair., 1980 Notes on Hyphomycetes - XXXIII *Stachybotrys spherospera* sp. nov. from S. Africa, Mycotaxon 10: 372-374.

Muller, E. & J.A. von Arx., 1973 Pyrenomycetes, Meliolales Coronophorales, Sphaeriales in the fungi (ed. Ainsworth et al.) Vol. IV A. 87-132. Acad. Pr. London.

Nag Raj T.R., & W.B. Kendrick., 1975 Genara Coelomycetarum-I Urohendersonia, Can. J. Bot. 49: 1853-1862.

Nilsson, S., 1964 Fresh water Hyphomycetes. Taxonomy, Morphology & Ecology. Symb. Bot. Upsaliense, 18: 1-130.

Nordbring-hertz, B. & M. Stalhammer-Carlemalm., 1978 Capture of nematodes by *Arthrobotrys oligospora,* an electron microscope study. Can. J. Bot. 56: 1297-1307.

Onions et al., 1981 Smith's Introduction to Industrial Mycology, Acad. pr. New York London.

Oudemans, C.A.J.A., 1885 Aanwinsten voor de flora mycologia Van nederland, Ned. Kruidk Archf. Ser. 2,4: 236-257.

Pacioni, G., 1985 The Mac Donald Encyclopedia of Mushrooms and Todestools. (English translation by Arnolodo Modadon Ed. S.P.A. Milan) Mac Donald & Co, London.

Padhya, R. & K.G. Mukerji., 1972 Fungi in the root region of *Rauwolfia serpentina* & *R.. canescens.* Indian Phytopath. 25: 104-107.

Palm, M.E., E.L. Stewart, & A.Y. Rossaman., 1979 *Pithomyces funiculosa* sp. nov. from fungal comb of Macrotermers ukezii in Swaziland. Mycotaxon 10: 246-250.

Persoon, C.H., 1794 Neuer Versuch einer Systematischen Eithei lung der Schwamme. Neues Mag. Bot. 1:63-128.

Persoon, C.H., 1801 Synopsis Methodica Fungorum Pp. 706, Gottingen.

Peterson, J.E., 1987 *Ganoderma* in Northern Europe, The mycologist, CAB. IMI. 62-67.

Peterson. R.H., 1971 The Genera *Gomphus* and *Gloecantharellus* in N. America, Verlag von J. Cramer.

Peterson, R.H., 1973 Aphyllophorales II, The Clavariod and Cantharelloid Basidiomycetes. (in the Fungi-Ainsworth et al. eds.) Acad. Pr. New York, London. Pp. 351-368.

Peterson, R.H. & K.W. Hughes., 1997 A new species *Pleurotus,* Mycologia 89(1): 193-180.

Pilat, A & O. Usak., 1961 Mushroom and other Fungi, London.

Pirozynski, K.A., 1962 *Circinotrichum* & *Gyrothrix,* Mycol. Pap. 84: 1-28.

Pirozynski, K.A. & C.S. Hodges., 1973 New Hyphomycetes from South Carolina. Can. J. bot. 51: 157-173.

Preuss, C.G.T., 1851 Uebersicht untersuchter pilze, besonders aus der Umegegand von Hoyers Werda. Linnaea 24: 99-153.

Punithalingam, E., 1970 Studies on Sphaeopsidales inculture. Mycol. Pap. 119: 1-24.

Ramakrishna, K. & C.V. Subramanian., 1931-1951 The Fungi of India a second supplement. J. Madras univ. 22(2): 163-182.

Rambelli, A. & D. Lunghini., 1979 *Chaetopsina* sp. from Tropical forest litter. Trans. Brit. mycol. Soc. 72: 491-493.

Rambelli, A. & S. Onofri., 1987 New species of *Kylindria* and *Xenokylindria* and notes on *Cylindrotrichum* (Hyphomycetes), Trans. Br. mycol. Soc.88:393-397.

Ramsbottom, J., 1951 Hand book of Larger Fungi, London.

Ramsbottom, J., 1953 Mushroms and Toadstools, Collins, New Naturalist.

Rao, D. & P.R. Rao., 1964 *Sporoschisma.* Berk & Br. from India. Mycopath. Mycol.Appl.24:81-84.

Rao, D. & P.R. Rao., 1966 Evolutionary trends in species of *Periconia*. Mycol. Appl. 28: 153-160.

Rao, D. & P.R. Rao., 1968 Studies on Tuberculaciaceae of Hyderabad-I. Sydowia 22: 170-172.

Rao,P.R. & D. Rao., 1964 Some helicosporae from Hyderabad-I Mycopath. Mycol. Appl.22: 47-54.

Rao, P.R. & D. Rao., 1964 The gunus *Periconia* from India. Mycopath.Mycol. Appl.22: 285-310.

Rao,P.R. & D. Rao., 1964 *Berkleasmium* Zobel from India. Mycopath. Mycol.Appl.22:311-314.

Rao, P.R. & D. Rao., 1964 Some allid dematiaceae dityosporae from Infia. Mycopath. Mycol.Appl. 23:23-28.

Rea, C., 1922 British Basidiomycetae Cambridge.

Reddy, S.S., S.M. Reddy & C. Manoharachary. 1975 A new species of *Tharoopama* Subram., Curr. Sci. 44: 719-720.

Reddy K.A., 1983 studies on Microfungi (Hyphomycetes) from Orissa, Ph.D. thesis 41-42. (unpublished).

Reddy, & Vasant Rao, 1984, *Bahugada* - a new Hyphomycete Curr.Sci., 54: 147-149.

Reynolds, D.R., 1971 Wall Structure of a bitunicate ascus. Planta. 98: 244-257.

Saccardo, P.A., 1886 Sylloge Fungorum Ominum Hecusque Cognitoru 4: 1-807.

Sarabhoy, A.K., D.K. Agarwal & J.L. Varshney., 1977-1981 Fungi of India, Associated Pub.Co. New Delhi.

Sathe, A.V. & S. Deshpande., 1980 Agaricales (Mushrooms) of S.W. India. in

3-parts Monograph-1. MACS Pune. Pp. 10-12.

Satya, A.V., & P.G., Rajyalaxmi, 1965, A new species of *Dieghtoniella* from Bopal, Curr.Sci., 34: 122-123.

Saxena, G. & N. Mittal., 1995 Trap formation of conidia of Nematode trapping *Monacrosporium* spp. Mycol. Res. 99: 839-840.

Seifert, K.A., 1985 A monograph of *Stilbella* and some allied Hyphomycetes. Stud. Myco. 27: 1-235.

Schaffer, R.L., 1957 *Volvariella* in N. America, Mycologia, 49: 545-589.

Schaffer, R.L., 1975 The major groups of Basidiomycetes, Mycologia, 67: 1-18.

Smith, G., 1962 Some new and interesting species Microfungi-III. Trans. Brit. mycol. Soc. 45: 387-394.

Sivanesan, A., 1984 The bitunicate Ascomycetes and their anamorphs. International Books & Periodical Suppl. Ser. New Delhi. Pp.1-701.

Sivanesan, A. & I.A.S. Gibson., 1976 CMI descriptions of pathogenic fungi & Bacteria. No. 481.

Soothill, E. & A. Fairhurst., 1978 The new Field Guide to Fungi. Michael Joseph, London.

Sree Ramulu, T. & A. Ramalingam., 1962 Notes on air borne *Tetraploa.* Curr. Sci. 31:121-122.

Sree Ramulu, T. & Ramalingham, 1962, Some observations *Deightoniella* Curr. Sci., 31:258-259.

Subramanian, C.V., 1952 Fungi Imperfecti from Madras-II Proc. Indian Acad.Sci.B36(4): 160-162.

Subramanian, C.V., 1955 Some species of *Periconia* from India. J. Indian Bot. Soc. 34: 339-361.

Subramanian, C.V., 1956 Hyphomycetes-I,II. J. Indian Bot. Soc.35: 53-91.

Subramanian, C.V., 1962 a The classification of the Hyphomycetes. Curr. Bull. Bot. Surv. India. 4(1-4): 249-259.

Subramanian, C.V., 1962 b A classification of the Hyphomycetes. Curr. Sci. 31: 409-411.

Subramanian, C.V., 1963 *Dactylella, Monacrosporium* and *Dactylina* J. Indian Bot. Soc. 42:291-300.

Subramanian, C.V., 1971 Hyphomycetes, an account of Indian species except Cercosporae. ICAR, New Delhi. Pp. 1-930.

Subramanian, C.V., 1977 Revision of Hyphomycetes-I Kavaka, 5: 93-98.

Subramanian, C.V., 1983 Hyphomycetes-Taxonomy and Biology. Acad. pr. London, New York. Pp. 1-502.

Subramanian, C.V. & K. Ramakrishnan., 1952-1956 List of Fungi of India. J. Madras Univ. 24(2): 327-421.

Sukapure, R.S. & M.J. Thirumalachar., 1963 Studies on *Cephalosporium* sp. from India-I. Mycologia, 55: 563-569.

Sutton, B.C., 1969 Forest Microfungi-III The Heterogenecity of *Pestalotia* de Not. Can. J. Bot. 48: 2083-2094.

Sutton, B.C., 1971 Conidium ontogeny in Pycnidial and acervular fungi, in Taxonomy of Fungi Imperfecti (ed. Kendrick) Pp. 263-278.

Sutton, B.C., 1973 Hyphomycetes from Manitoba and Saskatchewan, Canada. Mycol. Pap. 132: 1-143.

Sutton, B.C., 1980 The Coelomycetes-Fungi Imperfecti with Pycnidia, acervuli and Stromata. CMI, Kew, Surrey, England. Pp. 1-696.

Stahel, G., 1937 Banana leaf spot in Suriname caused by *Chloridium musae* nov spec and another related Banana disease, Trop. Agric. Trinidad 14: 42-44.

Sydow, H. et al., 1926-1972 Sylloge Fungorum vols. 24,25 & 26.

Talbot, P.H.B., 1971 The Priniciples of Fungal Taxonomy Pp. 274, Mac Millan.

Talbot, P.H.B., 1973 Aphyllophorales -I General characteristics; Thelephoroid and cupuloid Families, (in The Fungi vol. IV B Ainsworth et al.) Acad. Pr. New York London, 327-349.

Tandon, R.N. & Sudhirchandra., 1957-1962 Fungi of India, Bishen Singh Mahendrapal Singh New Delhi. Pp. 1-246.

Thind, K.S., 1977 The Myxomycetes of India ICAR, New Delhi.

Thind, K.S. & J.S. Dargan., 1978 Xylariaceae of India-IV. The genus *Daldinia* Kavaka 6: 15-24.

Tilak, S.T. & R. Rao., 1964 Fungi of India Sarada Pub. Aurangabad M.S. Pp.1-312.

Tubaki, K., 1958 Studies on the Japanese. Hyphomycetes-IV. J. Hattori Bot. Lab. 20: 142-145.

Tubaki, K., 1981 Hyphomycetes- Their perfect-imperfect connections, J. Cramer, Vaduz Pp. 1-181.

Tulasne, L.R., 1851 Fungi hypogaei Histoire et monographie des champignous hypoges Paris. (Microfiche).

Tulloch, M., 1972 The genus *Myrothecium* Tode ex Fr. Mycol. Pap. 13: 1-42.

Upadhyay, H.P., 1966 Soil fungi from North East Brazil-II, Mycopath. Mycol. Appl. 30: 276-282.

Upadhyay, H.P., 1967 Soil fungi from North East and North Brazil-IV, Nova Hedwigia 13: 227-234.

Varhese, K.M. & V.G. Rao., 1978 Two new setose Hyphomycetes from India. Bot. Notiser 131: 215-217.

Vasant Rao, 1983, Studies on hyphomycetes from S.India and monogrpah on *Satwalekara* Ph.D. thesis M.U. Aurangabad. (unpublished).

Vasant Rao, & G.S. de Hoog., 1975 Some notes on *Torula,* Persoonia 8: 199-206.

Vasant Rao, & G.S. de Hoog., 1986 New or critical hyphomycetes from India. Stud. Mycol. No. 28: 1-100.

Vasant Rao, & K.A., Reddy., 1978 Some new microfungi from India, Indian J. Mycol. Res. 16: 301-309.

Vasant Rao & K.A., Reddy, 1981 Two new Hyphomycetes. Indian J. Bot., 4(1) : 109-114.

Vasant Rao, K.A., Reddy, & G.S., de Hoog, 1984 *Heliocephala* - a new genus of dematiaceous hyphomycetes. Persoonia, 12:239-242.

Vasant Rao, & A. Satyanarayana., 1980 A note on *Tetraploa* Berk. & Br. Indian J. Bot. 3 (Suppl) 197-198.

Vasudeva, R.S., 1954 Fungi of India, ICAR, New Delhi.

Vasudeva, R.S., 1960 Fungi of India, ICAR, New Delhi. Pp. IX + 552.

von Arx, J.A., 1974 The Genera of Fungi sporulating in Pure culture. J. Cramer, Vaduz. Pp. 1-315.

Vuillemin, P., 1910 a Materiaux pour une classification rationelle des Fungi Imperfecti, C.R. Acad. Sci. Paris, 40: 882-884.

Vuillemin, P., 1910 b Les conidiophores. Bull. Soc. sci. Nancy. Ser. III, 11(2): 129-172.

Vuillemin, P., 1911 Les Aleurospores. Bull. Soc. sci. Nancy. Ser. III, 12: 151-175.

Wakefield, E.M. & G.R. Bisby., 1941 List of Hyphomycetes recorded for Britain. Trans. Brit. mycol. Soc. 25: 49-126.

Walting, R., 1970 Identification of larger Fungi, Hulton, Amersham.

Webster, J., 1980 Introduction to Fungi 2nd Ed. Cambridge Univ. Pr. London Cambridge.

Zucconi, L. & S. Onofri., 1989 *Gyrothrix ramosa* sp. nov. Mycol. Res.92(3): 380-382.

Zuck, R.K., 1946 Isolates between *Stachybotrys* and *Memnoniella,* Mycologia 38: 69-76.

VI Glossary

(Modified and adapted from Ellis 1976, Hawksworth et al. 1995)

Acerose	Needle-shaped.
Acroauxic	Growing and elongating at the apex.
Acrogenous	Borne at the tip.
Acropetal	Produced in succession towards the apex.
Acropleurogenous	Borne at the tip and along the sides.
Acuminate	Tapered to a point.
Allantoid	Sausage-shaped and slightly curved.
Ampulla	A vesicular swelling on a conidiophore, formed at the apex or at any point along its length
Ampulliform	Flask-shaped with the swollen part at the base.
Annellate	With ring-like markings on the wall showing where percurrent growth has taken place repeatedly.
Annellide	A holoblastic, closely annellate, percurrent conidiogenous cell.
Appendiculate	With one or more appendages (setulae).
Appressorium	An outgrowth of hypha which forms an organ of attachment.
Arthroconidium	A condium produced by a thallic conidiogenous cell; often cylindrical with truncate ends.
Basauxic	Growing and elongating at the base.
Basipetal	produced in succession towards the base.
Biconic	Cone-shaped at each end.
Blastic	Blown out, with marked enlargement of the conidium (conidiogenous cell) initial taking place before it is delimited by a septum.
Botryose	Clustered like grapes.

Caespitose	Aggregated in tufts.
Calyciform	Cup-shaped.
Capitate (hypha or seta)	A bulb-like swelling at the apex and resembling a drumstick.
Catenate	In chains.
Caudate	With a tail at the base.
Cerebriform	Convoluted like a brain.
Cheiroid	The shape of a hand roughly.
Cicatrized (Conidiogenous cell, Conidiophore)	Bearing scars.
Circinate	Coiled inwards.
Clavate	Club-shaped, thickened towards the apex.
Collaretter (Conidiogenous cell)	A small, expanded collar-like structure terminating or placed laterally on a phialide.
Conidiogenous cell	A cell which produces a conidium or several conidia.
Conidiophore	A hypha which bears a conidium or several conidia.
Conidium	Any asexual spore which when mature is liberated from a conidiophore or conidiogenous cell.
Conidiole	Small conidium
Corniculate	Having horn-like projections.
Cuneiform	Wedge-shaped.
Cylindrical	Round in cross section and of the same diameter throughout its length.
Cymbiform	Boat-shaped.
Denticle	A cylindrical or tapered, tooth-like projection on which a conidium is borne.
Denticulate (Conidiogenous cell, conidiophore)	Bearing denticles.

Determinate (Conidiogenous cell, conidiophore)	Ceasing to grow in length on the production of a terminal conidium or chain or conidia.
Dichotomous	Branching, often successively into two more or less equal arms.
Discrete (Conidiogenous cell)	Separate from the main axis or branches of a conidiophore and often having a distinctive shape.
Doliiform	Barrel-shaped.
Echinulate	Spiny
Endogenous	Produced entirely within a phialide.
Enteroblastic (Conidiogenous cell)	Blastic with the inner wall only or neither wall contributing towards the formation of the conidium wall.
Exogenous	Formed on the surface of a protoplasmic mass protruding from the end of a phialide.
Falcate	Sickle-shaped.
Falciform	Formed like a bill-hoock.
Fasciculate	Aggregated in bundles.
Filiform	Slender, thread-like.
Flexuous	Bent alternately in opposite directions.
Funiculose	Occurring in ropes or bundles.
Fusiform	Spindle-shaped, tapering at each end.
Geniculate	Bent like a knee, zig-zag
Helicoid	Form of a spiral.
Hilum	A scar showing the point of attachment.
Holoblastic (Conidiogenous cell)	Blastic with both the outer and inner walls contributing towards the formation of the conidium wall.
Hypha	A fungal thread or filament.
Hyphopodium	A short, lateral, sometimes lobed appressorium which adheres firmly to the surface of a host plant; a pale spot in the centre often indicates the point of origin of a narrow filament which penetrates the cuticle.

Infundibuliform	Funnel-shaped.
Integrated (Conidiogenous cell)	Incorporated in the main axis or branch of a conidiophore where it is either terminal or intercalary.
Intercalary	Between the apex and the base.
Lageniform	Like a long flask (a Florence flask).
Lenticular	Like a biconvex lens.
Limoniform	Lemon-shaped.
Lunate	Crescent-shaped.
Macronematous (Conidiophore)	Morphologically very different from a vegetative hypha and usually erect.
Meristematic (Conidiophore)	Increasing in length with the laying down of cross walls and elongation of cells.
Micronematous (Conidiophore)	Morphologically resembling a vegetative hypha but bearing conidia.
Monoblastic (Conidiophore)	Holoblastic and blowing out at only one point.
Mononematous (Conidiophore)	Composed of a single thread or filament.
Monophialidic (Conidiogenous cell)	Phialidic with only one opening.
Monopodial (Conidiophore)	With the axis continuing to grow at the apex always in the directions of previous growth; lateral branches produced behind the apex growing in a similar manner.
Monotretic (Conidiogenous cell)	Tretic with a single channel or pore.
Mother cell (Conidiophore)	The cell in a basauxic conidiophore from within which an extensible, usually conidiogenous filament arises; the apex of the externsible filament becomes incorporated in the first-formed conidium or a sterile cell and subsequent lateral conidia and sterile cells, when present, arise blastically in succesion towards the base.

Muriform	Longitudinally and transversely septate.
Mycelium	A mass or group of hyphae making up the thallus of a fungus.
Navicular	Boat-shaped.
Obclavate	The shape of a club upside down, thickened towards the base.
Obovoid	The shape of an egg upside down with the narrow end at the base.
Obpyriform	The shape of a pear upside down with the broad end at the base.
Obturbinate	The shape of a top upside down with the broad end at the base.
Obturbinate	The shape of a top upside down with the broad end at the base.
Obovoid	Egg shaped with the broad end at the base.
Pectinate	Resembling a comb.
Pedulous	Hanging down
Penicillate	A paint brush-like arrangement.
Percurrent (Conidiogenous cell, Conidiophore)	Growing straight on either through the open end left when the first conidium becomes detached or through an apical pore.
Phialide	A discrete or intergrated phialidic conidiogenous cell.
Phialidic (Conidiogenous cell)	Enteroblastic and producing conidia, usually in large numbers, in basipetal succession through one opening or several openings which are often provided with collaretters, and with neither the outer not inner wall contributing towrds the formation of hte conidium wall.
Phialoconidium	A conidium produced by a phialide.
Pleurogenous	Borne along the sides.
Polyblastic (Conidiogenous cell)	Holoblastic and blowing out at several points.
Polytretic (Conidiogenous cell)	Tretic with several channels or pores.

Prosenchymatous (Stroma)	Made up of elongated hyphal filaments still recognisable as such
Pseudoparenchyama-tous (stroma)	Resembling the parenchyma of flowering plants.
Pseudoseptate	With the appearance of having septa which do not reach acroos the conidium from wall to wall.
Pulvinate	Cushion-shaped.
Punctiform	Dot-like.
Pyriform	Pear-shaped with the broad end uppermost.
Reniform	Kidney-shaped.
Reticulate	Looking like a net.
Rostrate	Beaked or strongly attennuated at the apex.
Rugose	Coarsely wrinkled.
Rugulose	Finely wrinkled.
Schizogenous	Formed by cleavage or splitting.
Sclerotium	A frim mass of aggregated hyphae often with a rind of rather small, dark, thick-walled cells.
Semi-endogenous	Formd partly within a phialide.
Semi-macronematous (Conidiophore)	Intermediate in form between micronematous and macronematous, differing only slightly from avegetative hypha, often ascending but seldom erect.
Separating cells	Cells which separate conidia from each other or from conidigenous cells.
Seta	A bristle-like, often erect and thick-walled hypha.
Setiform (Conidiophore stipe or branch)	Having the form of a seta.

Setula	A delicate appendage arising from the surface of a conidium.
Sigmoid	S-shaped
Solitary	Arising singly at one point.
Spathulate	A spatula or spoon like.
Spinulose	Covered with little spines.
Sporodochium	A pulvinate stroma with closely packed, relatively short conidiophores covering its upper surface.
Stellate	Star-shaped.
Stipe (Conidiophore)	The main, stout axis or unbranched lower part.
Stomatopodium	An appressorium produced from a lateral branch above or in a stoma.
Stroma	An often cushion-like mas of fungal cells or closely interwoven hyphae.
Subulate	Awl-shaped.
Sympodial (Conidiogenous cell, Conidiophore)	Proliferating, the main axis elogating by growth of a succession of apices each of which develops behind and toone side of the previous apex where growth had ceased with the production of a conidium or conidia.
Synnematous (Conidiophore)	Broom-like, numerous threads or filaments tightly adpressed and in some cases fused along most of their length.
Thallic (Conidiogenous cell)	Conidium development from hyphae, without enlargement of the recognissable conidium initial, or when such enlargement does occur it takes place after the initial has been delimited by a septum or septa.
Thalloconidium	A conidium produced by a thallic conidiogenous cell.
Torulose	Swollen at intervals like beads in chain.
Torsive	Spirally twisted.

Tretic (Conidiogenous cell)	Apparently enteroblastic, protrusion of the inner wall taking place through one channel or several channels in the outer wall. The channels or pores, according to whcih way you view them can be seen quite easily with an ordinary compound microscope and although some recent elect4ron-microscope studies indicate that conidium development is not necessarily strictly enteroblastic their presence provides a character most useful to taxonomists.
Unciform	Hook-shaped.
Verrucose	Warted.
Verruculose	Finely warted.
Verticillate	Arranged in whorls.

KEY FOR IDENTIFICATION OF TAXA

I MYXOMYCOTINA :

Myxomycetes - True slime moulds.

1. Trophic phase free living............................ i (a)

2. Trophic phase not free living...................... i (b)

i (a) Spore mass dark coloured.
Peridium or capillitium calcarious .... Physarales.
Didymiaceae.. *Diderma* 16

i (b).Spore mass dark coloured.
Peridum or capillitium not calcarious......... Stemonitales
Stemonitaceae
Capillitium short.. *Comatricha* 8

II. BASIDIOMYCOTINA :

Basidiomycates - Mycelium septate, binucleate, clamp-connections, meiosporic, basidia present, doliipore septa.

1. Basidiomycetes.. *Agaricus*, etc.

2. Teliomycetes... Rusts - not included.

3. Ustomycetes... Smuts - not included.

(On basidial characters)

1. Teliomycetes
2. Hemi-basidiomycetes
3. Phragmo-basidiomycetes
4. Holo-basidiomycetes

} Not included fleshy, hymenium, besidiospores on basidial tips.

5. Agaricales
   Agaricaceae.. *Agaricus* 4

6. Ganodermatales
   Ganodermataceae.................................. *Ganoderma.* 17

7. Phallales
   Phallaceae.. *Phallus* 36

8. Poriales (Apphyllocphorales)
   Polyporaceae.. *Polyporus* 37

9. Sclerodermatales
   Sclerodermateaceae............................. *Scleroderma* 42

III. ASCOMYCOTINA :

Ascomycetes - Mycelium septate, binucleate, crozier, meiosporic, ascocarp, asci and ascospores. Ascocarp flask shaped with long neck, asco spores ooze from the ostiole (pore).

10. Ophiostomatales
    Ophiostomataceae................................ *Ophiostoma* 30

Ascocarp regular or irregular with ostiole (pore).

11. Xylariales
    Xylariaceae.. *Xylaria* 58

12. Daldinales
    Daldinaceae... *Daldinia.* 11

IV. DEUTEROMYCOTINA :

Deuteromycetes - Mitosporic, anamorphic, mycelium septate, forming conidiomata, (synnemata free, conidiophore, stromatic or sporodochial or acervular)

Synnemata present :

*	Synnemata compact or loose, at apex, polyphialidic, conidia guttulate, oval	*Paathramaya*	31
*	Synnemata compact or loose, polyblastic, conidia thin walled, oval or round.	*Phaeoisaria*	35
*	Synnemata compact with setae, conidia phialidic, sickle shaped or so....................	*Menisporopsis*	26
*	Synnemata compact with globose head, setae radiating, capitate, phialo-conidia......	*Saccardaea*	41
*	Synnemata with single head divided conidiogenous cells, conidiophores macronematous, conidia variable	*Tharoopama*	49
*	Synnemata absent, conidiophores macronematous..	*Periconiella*	33
*	Conidiophores phialidic / polyblastic Conidiophores not phialidic / polyblastic		
*	Setae branched, conidiophores at the base on mycelial mat..	*Gyrothrix*	19
*	Conidiophores on the setae..........................	*Gonytrichum*	18
*	Conidiophores macronematous with conidial head	*Periconia*	32
*	Conidiophores macronematous with out conidial head..	*Heliocephala*	21
*	Conidiophores macronematous, polyblastic, denticles large..........................	*Olpitrichum*	29
*	Denticles small	*Pseudobotrytis*	39

*	Denticles absent....................................	*Scolecobasidium*	43
*	Rounded conidia....................................	*Acrogenospora*	3
*	No denticles or very short conidia, fusiform..	*Latericonis*	23
*	Variable conidia	*Hansfordia*	20
		Ramichloridium	40
		Tritirachium	55
		Acrodontium	2
*	Conidiomata stromatic...............................	*Weisneriomyces*	57
*	Conidiophores macronematous, geniculate muriform....................................	*Bahugada*	6
*	Conidiophores micronematous, conidia slightly inflated curved..............................	*Cirrenaelia*	8
*	Conidiaphores macronematous not geniculate Percurrent conidiophores	*Pithomyces*	38
*	Not percurrent conidiophores		
*	Armed conidia	*Tetraploa*	48
*	Not armed conidia..................................	*Trimmatostroma*	54
*	Conidiophores cupulate or globular........	*Torula*	51
*	Conidiophores elongated percurrent monocrogenous........................	*Melanocephala*	24
*	Conidiophores elongated, conidiogenous cells in verticles pleurogenous	*Moorella*	28
*	Conidiophores in pycnidia........................	*Pestalotia*	34
*	Conidiophores enteroblastic, phialidic, tretic..	*Dendryphiopsis*	13

*	Conidiophores monophialidic, phragmo-conidia in chains....................	*Sporoschisma*	45
	1-celled conidia solitary..........................	*Acremonium*	
	2-celled conidia solitary..........................	*Arthrobotrys*	5
*	Conidiophores polyphialidic, conidia cylindrical..................................	*Cylindrocladium*	10
*	Conidiophores with phialides on apex..	*Memnoniella*	25
		Stachybotrys	46
*	Conidiophores with phialides all not in verticels...	*Stachylidium*	47
*	Along in verticels	*Verticillium*	56
*	Conidiophores with phialides, conidia in slimy mass..	*Trichoderma*	53
		Tolypocladium	52
*	Conidiophores basauxic, 1-celled............	*Spegazzinia*	44
*	Conidiophores basauxic, many celled.....	*Dictyoarthrinium*	14
*	Fungi mitosporic but not with definite conidiogenesis...	Mycelia Sterilia	
*	Fungi mitosporic but with definite conidiogenesis...	(Discussed above)	
*	Conidiomata dendroid, propagules disc-shaped..	*Tretopileus*	51
*	Propagules not disc-shaped	*Kinnerasaniomyces*	21

www.ingramcontent.com/pod-product-compliance
Ingram Content Group UK Ltd.
Pitfield, Milton Keynes, MK11 3LW, UK
UKHW021327190726
13845UKWH00004B/54